R00162 97047

D1563774

CHICAGO PUBLIC LIBRARY
HAROLD WASHINGTON LIBRARY CENTER
R0016297047

Principles of Electrochemical Machining

Principles of Electrochemical Machining

J. A. McGEOUGH
Department of Engineering
University of Aberdeen

CHAPMAN AND HALL
London
A HALSTED PRESS BOOK
JOHN WILEY & SONS
New York

First published 1974
by Chapman and Hall Ltd
11 New Fetter Lane, London EC4P 4EE

Printed in Great Britain by
William Clowes & Sons Limited
London, Colchester and Beccles

Library of Congress Cataloging in Publication Data

McGeough, J A
Principles of electrochemical machining.

Includes bibliographies.
1. Electrochemical cutting. I. Title.
TJ1191.M24 671.3′5 74–2780
ISBN 0–470–58413–0

Preface

The nature of the electrochemical machining process has meant that the efforts made towards understanding its principles lie in several different fields. Although increasing light on the subject continues to be shed by studies in these separate disciplines, the engineer still carries the problem of making judgements which embrace the entire process. He should, therefore, be conversant with the appropriate features of all the relevant fields.

The purpose of this book is to contribute towards overcoming this problem by describing and correlating those aspects of the relevant fields which have proved useful in engineering studies of electrochemical machining. To that end, the first three chapters are devoted to a discussion of the developments which have led to the need for electrochemical machining, and the fluid dynamic and electrochemical principles on which the process is based. It is recognised that specialists in these two main fields may regard the treatment of their subjects as elementary, but they will probably accept that these topics are not necessarily closely related, and that acquaintance with both is necessary in electrochemical machining work. The contents of each of these chapters can be studied in greater detail with the help of the bibliography given at the end of the chapters. In the fourth chapter, the characteristics of metals electrochemically machined in different electrolytes are examined in the light of established and recent theories. Chapter 5 deals with the dynamics and kinematics of the process. The sixth and seventh chapters are concerned with the fundamental problems of the process, namely, the prediction of the change of anode shape with machining time, and the design of cathode tool shapes to machine an anode workpiece to a specified form. In the final chapter, the principles of the process described in the preceding chapters are illustrated by examples from industrial practice. Throughout the book, calculations are carried out at appropriate stages to give the reader a grasp of the physical dimensions of the quantities involved in the process. Although SI units have been preferred, the character of the subject has not easily lent itself towards the exclusive use of this system. Accordingly, some quantities are given in other metric units which are in common use.

A large number of people have to be thanked who, in many different ways, have contributed towards the writing of this book. First, I should like to thank Professor G. D. S. MacLellan for a patient introduction to the subject. Appreciation must also be expressed to Professor L. Bass, who spent much time giving explanations of electrochemical phenomena; and I am indebted to Professor T. M. Charlton whose advice and wholehearted support have eased so much my task. Mr. P. Lawrence, Drs. D. G. Lovering, N. S. Mair, G. D. Mathew and H. Rasmussen, and Messrs. D. Stewart and A. F. Stronach all read sections of the book and were very helpful in making suggestions for its improvement. Calculations were checked by Mr. S. H. F. Lai. Mr. R. Penny prepared the diagrams with the help of Messrs. D. Bain and I. Burt. Two British manufacturers of ECM equipment, Healy of Leicester Ltd., and Herbert Machine Tools Ltd., Lutterworth, kindly supplied the photographs of industrial applications shown in Chapter 8. The manuscript was typed skilfully and patiently by Miss A. Campbell, with a useful contribution from Mrs. J. Fogarty. I am also grateful to my colleagues in the Engineering Department at Aberdeen University who have done much to lighten my load during the preparation of the book, and to Mr. R. Stileman of Chapman and Hall Ltd. for his consideration.

Finally, I wish to express my greatest thanks to my wife who has remained a steadfast source of encouragement throughout the exercise.

J. A. McGeough
Aberdeen

Contents

Principal Notation

a	Tafel constant $[= (-2.303RT/z\alpha F) \log J_0]$
a	length of mesh 'arm' (Chapter 7)
a, $[a]$	activity, activity term
a_n	Fourier coefficient (Chapter 6)
A	electrode area (Chapter 1)
A	atomic weight
A	logarithmic throwing index (Chapter 4)
A	cross-section area of flow channel (Chapter 2)
A	dimensionless process quantity $(= \rho_0 e_g / \sigma \rho_g e_a)$ (Chapter 5)
b	Tafel constant $(= 2.303RT/z\alpha F)$
b	width of electrode
b_n	Fourier coefficient (Chapter 6)
B	dimensionless process quantity $(= \zeta T_r)$ (Chapter 5)
$c(y)$	ion concentration (particle numbers per unit volume of solution) (Chapter 3)
c_e	specific heat of electrolyte
c_n	Fourier coefficient (Chapter 6)
C	concentration of electrolyte solution
C	wetted perimeter (Chapter 2)
C_b	bulk concentration of reacting species (Chapters 3, 4, 5)
C_H	hydrogen ion concentration
C_S	interfacial concentration of dissolution products
C	dimensionless process quantity $[= (2/\sigma)(e_g/e_a)(RT_0/U_0^2)]$ (Chapter 5)
d	pipe diameter
d_h	hydraulic diameter
d_n	Fourier coefficient (Chapter 6)

D	diffusion coefficient
e	electron charge
e_a	electrochemical equivalent of anode metal
e_g	electrochemical equivalent of gas
E	reversible electrode potential
$\mathbf{E}$	electric field (Chapter 7)
E_δ	electric field across diffusion layer (Chapter 3)
E_0	normal electrode potential (Chapter 3)
f	activity coefficient (Chapter 3)
f	coefficient of frictional resistance (Chapter 2)
f	electrode (cathode) feed-rate
$f(J)$	arbitrary current density-dependent function of cathode overpotential (Chapters 4, 5, 6)
F	Faraday's constant (96 500 C)
$\mathbf{F}$	gravitational force (Chapter 2)
$g(J)$	arbitrary current density-dependent function of anode overpotential (Chapter 6)
h	inter-electrode gap width
$h_a(x)$	local width of layer of anodic products (Chapter 5)
h_e	equilibrium gap width
$h_{el}(x)$	local width of central layer of pure electrolyte (Chapter 5)
$h_g(x)$	local width of layer of hydrogen gas bubbles (Chapter 5)
h_0	equilibrium gap width at gap inlet
h_∞	total overcut (Chapter 7)
$h(t)$	width of gap at machining time, t
$h(0)$	width of gap at start of machining, $t = 0$
H_g	volume of hydrogen produced per coulomb (Chapter 4)
$\mathbf{i}$	unit vector in x direction (Chapter 2)
i	square root of (-1) (Chapter 7)
(i, j)	general coordinates of grid points (Chapter 7)
I	current
$\mathbf{j}$	unit vector in y direction (Chapter 2)
j	number flux of ion particles (Chapter 3)
J	current density
J_l	limiting current density (Chapters 3, 4, 5)
J_0	exchange current density
J_p	passivation current density
$\bar{J}$	a defined average current density, consequence of overpotential (Chapters 4, 6)
k	Boltzmann's constant

k	wave number
$\mathbf{k}$	unit vector in z direction (Chapter 2)
K_0	pressure loss coefficient for rectilinear gap without ECM (Chapter 5)
K_s	activity solubility product (Chapter 3)
l	characteristic length for flow path (Chapter 2)
l	half-width of cathode (Chapter 7)
L	electrode length
L	linear ratio (Chapter 4)
m	mass of metal removed, or deposited
$\dot{m}$	rate of metal removal
m_a	local mass flux rate for anodic products (mass per unit time per unit area) (Chapter 5)
m_g	local mass flux rate for hydrogen gas (Chapter 5)
M	metal distribution ratio (Chapter 4)
M	machining parameter ($= A\kappa_e/z\rho_a F$)
M	frictional pressure drop multiplier for two-phase flow
n	number of particles with energy to cross interface (Chapter 3)
n_e	electrochemical valency
n_t	total number of particles (Chapter 3)
N	Avogadro's number (6×10^{23})
Nu	Nusselt number
Nu_p	Nusselt number associated with passivation
p	pressure
$p(t)$	mean gap between electrodes (Chapters 4, 5, 6, 7)
p_e	mean equilibrium gap (Chapters 4, 6)
$p(0)$	mean gap between electrodes at time $t = 0$ (Chapter 6)
pH	negative logarithm of hydrogen ion concentration
$\mathbf{P}$	boundary force (Chapter 2)
Q	volumetric flow-rate
$\mathbf{r}_a$	vector function of anode position
$\dot{\mathbf{r}}_a$	dissolution rate of anode (Chapters 6, 7)
R	Gas Constant
R	radius of pipe (Chapter 2)
R	residual (Chapter 7)
R	resistance (Chapter 1)
Re	Reynolds number
S	dimensionless process quantity ($= \rho_0 h_0 U_0/L\rho_a f$) (Chapter 5)

Sc	Schmidt number ($= \nu/D$)
t	time of machining
t_+, t_-	cationic, anionic transport number (Chapter 3)
T	temperature
T	throwing power (Chapter 4)
T_b	boiling temperature
T_r	reference temperature ($= V/e_a c_e$) (Chapter 5)
u	electrolyte velocity in x direction
u_f	friction velocity
u_+	cationic mobility (Chapter 3)
u_-	anionic mobility (Chapter 3)
u_{max}	maximum electrolyte velocity
$\bar{u}$	mean electrolyte velocity
U	mainstream electrolyte velocity
U	electrolyte velocity vector (Chapter 2)
U_a	local velocity of layer of anodic products (Chapter 5)
U_e	local velocity of layer of pure electrolyte (Chapter 5)
U_g	local velocity of layer of gas (Chapter 5)
v	electrolyte velocity in y direction
$\dot{v}$	volumetric rate of metal removal
V	applied potential difference (voltage)
V_{rms}	applied voltage (alternating current conditions) (Chapter 5)
w	electrolyte velocity in z direction (Chapter 2)
W	uninsulated land width (Chapter 7)
$W_{1,2}$	energy required to pass interface from metal to solution (subscript 1) and vice versa (subscript 2)
x	coordinate distance along electrode length
x	number of cations (Chapter 3)
x_E	entry length (Chapter 2)
(x, y, z)	coordinate system
X_i	percentage by weight of element i (Chapter 4)
y	coordinate distance from cathode to anode, normal to x direction
y	number of anions (Chapter 3)
z	transverse direction to (x, y) axes
z	valency
z_+	charge on cations (Chapter 3)
z_-	charge on anions (Chapter 3)
α	degree of dissociation (Chapter 3)

α fraction of overpotential contributing towards dissolution

α void fraction

α average cathode overpotential function $[= f(\bar{J})]$ (Chapters 4, 5

β cathode overpotential function $[= \partial f(J)/\partial J]$ (Chapters 4, 6, 7

γ dimensionless process constant $(= h_0 U_0/Lf)$ (Chapter 5)

γ anode overpotential function $[= g(J)]$ (Chapter 6)

Γ current efficiency (Chapter 5)

δ diffusion layer thickness

$\bar{\delta}$ average value of diffusion layer thickness

δ_d displacement thickness (Chapter 2)

δ_g width of layer containing electrolyte and gas bubbles

δ_0 thickness of laminar boundary layer

δ_l thickness of viscous sub-layer

δ_t thickness of turbulent boundary layer

ΔG free energy change (Chapter 3)

ΔV sum of overpotentials and reversible potentials at both electrodes

ϵ_{adm} admissible height of surface projections (Chapter 2)

$\epsilon(t)$ amplitude (height) of surface irregularities

$\epsilon(0)$ initial value of amplitude of surface irregularities $(t = 0)$

ζ temperature coefficient for electrolyte conductivity

η_a activation overpotential

η_{conc} concentration overpotential

η_r resistance overpotential

θ momentum thickness (Chapter 2)

θ angle between normal to anode boundary and direction of cathode movement (Chapter 7)

κ_e electrolyte conductivity

κ_0 electrolyte conductivity at gap inlet

κ_m mean conductivity of electrolyte–gas medium

2λ fundamental wavelength of anode irregularities

λ_+ defined quantity $(= Fu_+)$ (Chapter 3)

λ_- defined quantity $(= Fu_-)$ (Chapter 3)

Λ_c equivalent conductivity (Chapter 3)

Λ_m molar conductivity

Λ_0 equivalent conductivity at very low limiting concentration

μ absolute viscosity (Chapter 2)

μ dimensionless overpotential parameter $(= \beta\kappa_e/\lambda)$

ν kinematic viscosity

(ξ, η) coordinate system in ζ-plane (Chapter 7)

ρ	resistivity (Chapter 1)
ρ_a	anode metal density
ρ_e	electrolyte density
ρ_g	gas density
σ	slip ratio ($= U_g/U_e$) (Chapter 5)
σ	dimensionless configuration parameter ($= p/\lambda$)
τ	anode overpotential function [$= \partial g(J)/\partial J$] (Chapter 6)
τ	machining constant (Chapter 7)
τ	shearing stress (Chapter 2)
τ_0	shearing stress at wall (Chapter 2)
ϕ	potential
$\partial\phi/\partial n$	normal component of electric field at anode surface
$\partial\phi/\partial s$	defined potential gradient along edge of conducting paper (Chapter 7)
ψ	thermionic work function (Chapter 3)
ψ	nondimensional process function $\{= fh(0)/[A\kappa_e(V - \Delta V)/zF\rho_a]\}$ (Chapter 5)
$(\psi_1 - \psi_2)$	contact potential difference (Chapter 3)
ω	overpotential parameter (Chapters 4, 6)
Ω	slip ratio ($= U_a/U_e$) (Chapter 5)
Ω_1, Ω_2	characteristic parameter for equilibrium dissolution, deposition reaction (Chapter 3)

Subscripts

a	property of anode
c	property of cathode
e	property of electrolyte
(h_e	equilibrium gap width)
g	property of gas
m	property of metal (Chapter 3)
0	condition at gap inlet
+, −	property of positive, negative electrolyte ions (Chapter 3)

(Other subscripts defined in text)

Superscripts

0	dimensionless process variable
*	dimensionless quantity (Chapter 5)

CHAPTER ONE

Introduction

Michael Faraday's early metallurgical researches, from 1818 to 1824, anticipated the developments which have led to the widespread use today of alloy steels. Of the innumerable applications of these materials, their service as cutting tools in machining practice has been recognised from the outset, and much effort has been expended to improve their performance. The aim has always been to yield higher rates of machining and to tackle harder metals which are developed (on the principle that the tool material must be harder than that of the workpiece which is to be machined). To that end, various heat-treatments and compositions of tool materials and

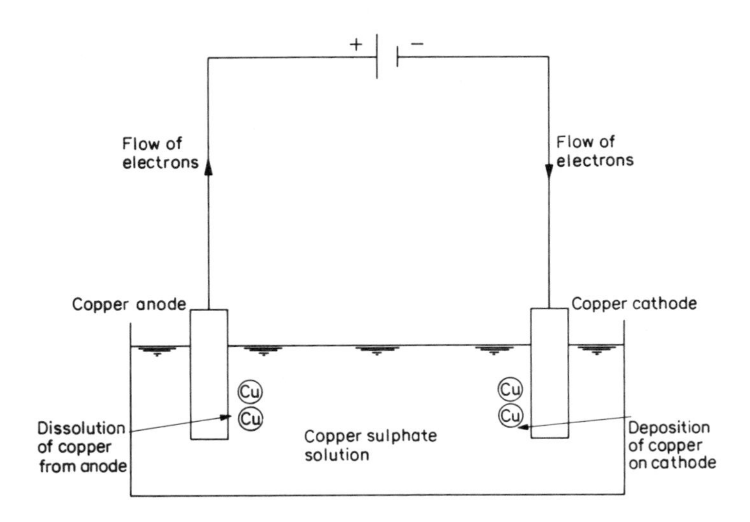

Fig. 1.1 Electrolysis of copper sulphate solution

formations of tools continue to be tried. Much progress has been made, but in recent years some alloys have been produced which are exceedingly difficult to machine. These have been prepared to meet a demand for very high-strength, heat resistant materials which, moreover, often have to take a complex shape. The evolution of suitable tooling has not kept pace with these advances, and accordingly, a search has had to be made for alternative methods of machining.

Electrochemical machining (ECM) has been developed initially to machine these alloys, although any metal can be so machined. Its basis is the phenomenon of electrolysis, whose laws were established by Faraday in 1833, and which, with his electrical studies, was mainly responsible for diverting his attention from his work on metals.

1.1 Electrolysis

Electrolysis is the name given to the chemical process which occurs, for example, when an electric current is passed between two conductors dipped into a liquid solution. A typical example is that of two copper wires connected to a source of direct current and immersed in a solution of copper sulphate in water, as shown in Fig. 1.1. An ammeter, placed in the circuit, will register a flow of current; from this indication, the electric circuit can be deduced to be complete. A significant conclusion that can be made from an experiment of this sort is that the copper sulphate solution obviously has the property that it can conduct electricity. Such a solution is termed an *electrolyte*. The wires are called *electrodes*, the one with positive polarity being the *anode*, and the one with negative polarity the *cathode*. The system of electrodes and electrolyte is referred to as the *electrolytic cell*, whilst the chemical reactions which occur at the electrodes are called the *anodic* or *cathodic reactions* or *processes*.

Electrolytes are different from metallic conductors of electricity in that the current is carried not by electrons but by atoms, or groups of atoms, which have either lost or gained electrons, thus acquiring either positive or negative charges. Such atoms are called *ions*. Ions which carry positive charges move through the electrolyte in the direction of the positive current, that is, towards the cathode, and are called *cations*. Similarly, the negatively charged ions travel towards the anode and are called *anions*. The movement of the ions is accompanied by the flow of electrons in the opposite sense outside the cell, as shown also in Fig. 1.1, and both actions are a consequence of the applied potential difference (or voltage) from the electric source.

A cation reaching the cathode is neutralised, or discharged, by the negative electrons on the cathode. Since the cation is usually the positively charged atom of a metal (in the above case, copper) the result of this reaction is the deposition of metal (copper) atoms. For copper, the reaction may be written

$$Cu^{++} + 2e \rightarrow Cu$$

where e is one electron.

To maintain the cathodic reaction, electrons are required to pass round the external circuit. These are obtained from the atoms of the metal anode, and these atoms thus become the positively charged cations which pass into solution. In this case, the reaction is the reverse of the cathodic. For the above example, it is

$$Cu \rightarrow Cu^{++} + 2e$$

The electrolyte in its bulk must be electrically neutral; that is, there must be equal numbers of opposite charges within it, and thus there must be equal amounts of reaction at both electrodes. Therefore, in the electrolysis of copper sulphate solution with copper electrodes, the overall cell reaction is simply the transfer of copper metal from the anode to the cathode. When the wires are weighed at the end of the experiment, the anodic wire will be found to have lost weight, whilst the cathodic wire will have increased in weight by an amount equal to that lost by the other wire.

These results are embodied in Faraday's two laws of electrolysis:

(i) the amount of any substance dissolved or deposited is directly proportional to the amount of electricity which has flowed;
(ii) the amounts of different substances deposited or dissolved by the same quantity of electricity are proportional to their chemical equivalent weights.

The two laws may be combined to give the equation

$$m = \frac{AIt}{zF} \tag{1.1}$$

where m is the mass dissolved from, or deposited upon, the metal by a current I passed for time t. The reacting ions have atomic weight A and valency z, the quantity A/z being the chemical equivalent. F is a universal constant known as the Faraday (or Faraday's constant). It is the amount of electric charge necessary to liberate one gram-equivalent (A/z) of an ion in electrolysis. Its commonly accepted value is 96 500 C.

When the product It is unity, i.e. one coulomb of charge has been passed, the value, $m = A/zF$, so defined is the *electrochemical equivalent* of the metal, e_a. For example, copper has an atomic weight of 63·57. When it is univalent, its equivalent weight is 63·57 g; when the metal is divalent, the equivalent weight is 31·78 g. The corresponding electrochemical equivalents are respectively 66×10^{-5} g/C and 33×10^{-5} g/C. Iron has an atomic weight of 55·85. In its divalent form this metal has an electrochemical equivalent of 29×10^{-5} g/C; trivalent iron has an electrochemical equivalent of 19×10^{-5} g/C.

A popular application of electrolysis is the electroplating process in which metal coatings are deposited upon the surface of a cathodically polarised metal. Current densities used are roughly 2×10^{-2} A/cm^2 and the thickness of the coatings is seldom more than about 10 μm. (Current density here is regarded as the current used over the area under treatment.) An example of an anodic dissolution operation is electropolishing. Here, the item which is to be polished is made the anode in an electrolytic cell. Irregularities on its surface are dissolved preferentially so that, on their removal,

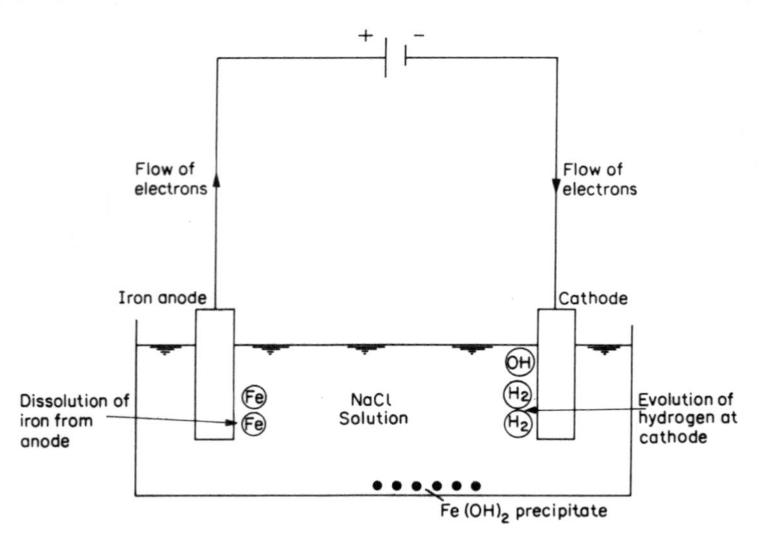

Fig. 1.2 Electrolytic dissolution of iron

the surface becomes flat and polished. A typical current density in this operation would be 10^{-1} A/cm^2, and polishing is usually achieved on the removal of irregularities as small as 10^{-2} μm. With both electroplating and electropolishing, the electrolyte is either in motion at low velocities or unstirred.

ECM is similar to electropolishing in that it also is an anodic dissolution process. But the rates of metal removal offered by the polishing process are considerably less than those needed in metal machining practice. To find how ECM meets these requirements and, moreover, how it is used to shape metals, we study first another type of electrolysis, namely, that arising from iron in aqueous sodium chloride (Fig. 1.2).

When a potential difference is applied across the electrodes, several possible reactions can occur at the anode and cathode. Certain reactions, however, are more likely to arise than others; this preference will be explained in Chapter 3 in terms of the *energy* that is available for each reaction. In the present example, the probable anodic reaction is dissolution of iron, e.g.

$$Fe \rightarrow Fe^{++} + 2e$$

At the cathode, the reaction is likely to be the generation of hydrogen gas and the production of hydroxyl ions:

$$2H_2O + 2e \rightarrow H_2 + 2OH^-$$

The outcome of these electrochemical reactions is that the metal ions combine with the hydroxyl ions to precipitate out as iron hydroxide, so that the net reaction is

$$Fe + 2H_2O \rightarrow Fe(OH)_2 + H_2$$

(Note that the ferrous hydroxide may react further with water and oxygen to form ferric hydroxide:

$$4Fe(OH)_2 + 2H_2O + O_2 \rightarrow 4Fe(OH)_3$$

although it is stressed that this reaction, too, does not form part of the electrolysis.)

With this metal–electrolyte combination, the electrolysis has involved the dissolution of iron from the anode, and the generation of hydrogen at the cathode, *no other action taking place at the electrodes.*

Certain observations relevant to ECM can be made at this stage:

(i) Since the anode metal dissolves electrochemically, its rate of dissolution (or machining) depends only upon the atomic weight A and valency z of ions produced, the current I which is passed, and the time t for which the current passes. The dissolution rate is not influenced by the hardness or other characteristics of the metal;

(ii) since only hydrogen gas is evolved at the cathode, the shape of that electrode remains unaltered during the electrolysis. This feature will be shown later to be most relevant in the use of ECM as a metal shaping process.

1.2 Development of the characteristics of ECM

These two aspects can be developed further by use of Equation (1.1). From that equation, and since $m = v\rho_a$, where v is the corresponding volume and ρ_a the density of the anode metal, the volumetric removal rate of anodic metal $\dot{v}$ is

$$\dot{v} = \frac{AI}{zF\rho_a} \tag{1.2}$$

Suppose that a machining operation has to be carried out on an iron workpiece at a typical rate, say, $0{\cdot}026 \times 10^{-6}$ m^3/s. For this removal rate to be achieved by ECM, the current in the cell must be about 700 A [on substitution in Equation (1.2) of the values $A/zF = 29 \times 10^{-5}$ g/C and $\rho_a = 7.86$ g/cm^3 for iron]. Currents used in ECM are of this magnitude, and indeed they are often higher, by as much as an order of magnitude. The corresponding average current densities are typically 50 to 150 A/cm^2.

The means by which these high current densities are obtained can be understood from an examination of the other characteristics of an ECM cell, in particular, the electrolyte conductivity and the inter-electrode gap width. These parameters are related to the current through Ohm's law, which states that the current I flowing in a conductor is directly proportional to the applied voltage V:

$$V = IR \tag{1.3}$$

In the simple expression (1.3), R is the resistance of the conductor. The experiments on electrolysis, described above, demonstrate that electrolytes are also conductors of electricity. Ohm's law also applies to this type of conductor, although the resistance of electrolytes may amount to hundreds of ohms.

Now, the resistance R of a uniform conductor is directly proportional to its length h, and inversely proportional to its cross-sectional area A. Thus

$$R = \frac{h\rho}{A} \tag{1.4}$$

where ρ is the constant of proportionality. If the conductor is a cube of side 10 mm, then $R = \rho$; ρ is termed the *specific resistance* or *resistivity* of the conductor. The reciprocal of the specific resistance is the *specific conductivity*, often denoted by the symbol κ.

If Equations (1.3) and (1.4) are combined, the following relationships are derived between the average current density, current, surface area to be machined, applied potential difference, gap width, and electrolyte conductivity, these quantities being denoted by the respective symbols J, I, A, V, h, and κ_e:

$$J = \frac{I}{A} = \frac{\kappa_e V}{h} \tag{1.5}$$

It has been pointed out that in practice J is often about 50 A/cm^2. To obtain a current density of this magnitude, a cell could be devised with high values for κ_e and V and low values for h. Even for strong electrolytes, however, κ_e is small. If the current is high, power requirements, amongst other considerations, restrict the use of high voltages, and, in practice, the voltage is usually about 10 to 20 V. If values of 0·2 ohm^{-1} cm^{-1} and 10 V are taken for κ_e and V respectively, then for J to be 50 A/cm^2 the gap h must be 0·4 mm. It will be shown later that a gap of this size is also necessary for accurate shaping of the anode. As dissolution of the anode proceeds, this gap is maintained by mechanical movement of one electrode, say the cathode, towards the other. To maintain the gap of 0·4 mm, a cathode feed-rate about 0·02 mm/s would be needed, the values given above for the other process variables being retained.

The accumulation within the small machining gap of the metallic and gaseous products of the electrolysis is undesirable. If the growth were left uncontrolled, eventually a short circuit would occur between the two electrodes. To avoid this crisis, the electrolyte is pumped through the inter-electrode gap so that the products of the electrolysis are carried away. The forced movement of the electrolyte is essential also in diminishing the effects of electrical heating of the electrolyte, due to the passage of current, and of hydrogen gas, which respectively increase and decrease the effective conductivity. These matters will be discussed in greater detail later, but at this stage, the Joule heating effect provides a simple, convenient way of estimating a typical electrolyte velocity. Without forced agitation to control the increase in the electrolyte temperature, boiling will eventually occur in the gap. If all the heat caused by the passage of current remains in the electrolyte, the temperature increase δT in a length δx of gap is, from Joule's and Ohm's laws,

$$\delta T = \frac{J^2 \, \delta x}{\kappa_e \rho_e c_e U} \tag{1.6}$$

where U is the electrolyte velocity, ρ_e the electrolyte density, and c_e its specific heat.

If, for simplicity, the increase with temperature of the electrolyte conductivity is neglected, integration of Equation (1.6) yields

$$U = \frac{J^2 L}{\kappa_e \rho_e c_e \Delta T} \tag{1.7}$$

where L is the electrode length and ΔT is the temperature difference of the electrolyte between points at inlet and outlet to the gap.

Consider the typical values, $J = 50$ A/cm², $L = 10^2$ mm, $\kappa_e = 0{\cdot}2$ ohm^{-1} cm^{-1}, $\rho_e = 1{\cdot}1$ g/cm³, $c_e = 4{\cdot}18$ J g^{-1} °deg C^{-1}. Suppose, too, that ΔT must be kept to 75°C to avoid boiling at the exit point, the inlet temperature being, say, 25°C. From Equation (1.7), the velocity to maintain this condition is calculated to be about 3·6 m/s. Velocities of the electrolyte solution through the gap in ECM usually range from about 3 to 30 m/s. The pressures required to achieve these velocities will be calculated in the next chapter.

1.3 Basic working principles

ECM has been founded on the principles outlined in Sections 1.1 and 1.2. As shown in Fig. 1.3, the workpiece and tool are made the anode and cathode, respectively, of an electrolytic cell, and a potential difference, usually fixed at about 10 V, is applied across them. A suitable electrolyte (e.g. aqueous NaCl solution) is chosen so that the cathode shape remains unchanged during electrolysis. The electrolyte, whose conductivity is about 0·2 ohm^{-1} cm^{-1}, is also

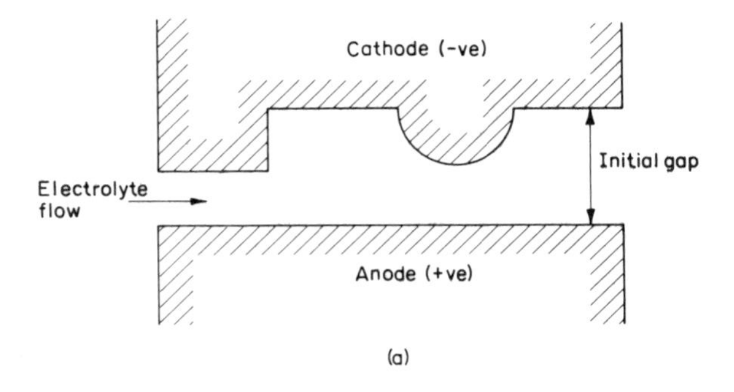

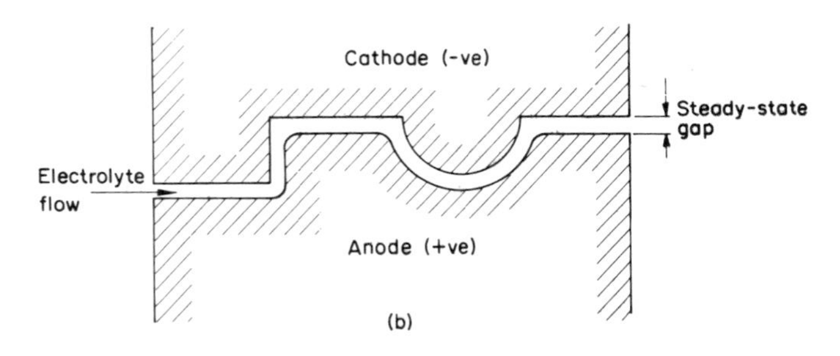

Fig. 1.3 (a) Initial electrode configuration for ECM. (b) Final electrode configuration for ECM

pumped at a rate, roughly 3 to 30 m/s, through the gap between the electrodes to remove the products of machining and to diminish unwanted effects, such as those that arise with cathodic gas generation and electrical heating. The rate at which metal is then removed from the anode is approximately in inverse proportion to the distance between the electrodes. As machining proceeds, and with the simultaneous movement of the cathode at a typical rate, say 0·02 mm/s, towards the anode, the gap width along the electrode length will gradually tend to a steady-state value. Under these conditions, a shape, roughly complementary to that of the cathode, will be reproduced on the anode. A typical gap width then should be about 0·4 mm and the average current density should be of the order of 50 to 150 A/cm^2. Moreover, if a complicated shape is to be formed on a workpiece of a hard material, the complementary shape can first be produced on a cathode of softer metal, and that cathode can then be used to machine electrochemically the workpiece. In short, the main advantages of ECM are:

(i) the rate of metal machining does not depend on the hardness of the metal;
(ii) complicated shapes can be machined on hard metals;
(iii) there is no tool wear.

Bibliography

Faraday, M., *Experimental Researches in Electricity*, Vols. I, II, III, reprinted by Dover, New York (1965).

Glasstone, S., *The Elements of Physical Chemistry*, Macmillan, London (1950).

Gusseff, W., British Patent No. 335 003.

Hadfield, R. A., *Faraday and His Metallurgical Researches,* Chapman and Hall, London (1931).

CHAPTER TWO

Basic Fluid Dynamics

In the first chapter, the need for electrolyte flow in ECM was established. An understanding of the fluid dynamics involved in the process can be obtained from the analysis of five basic equations, which describe the likely patterns of the flow. In particular, these equations provide a guide to the significance of flow velocity profiles. From such studies the pressures can be estimated that are required to pump the electrolyte at specific rates down the gap between the two electrodes, and useful information can be obtained about the variation in electrolyte velocity *across* the gap.

In the treatment of such matters in this chapter emphasis is given to flow along rectangular channels and circular pipes since these flow channels are common in ECM. Nevertheless, the principles which are discussed are applicable also to other flow configurations, such as those arising with the rotating disc. The use of that type of configuration in both fluid dynamics and electrochemical (including ECM) studies is well known [1, 2].

2.1 Basic assumptions and definitions

Investigations concerned with real, as distinct from perfect, fluid motion rely heavily on two assumptions. The first is that, wherever the fluid is in contact with a solid boundary, there is no motion or slip, relative to that boundary, of the fluid particles adjacent to it. In the second assumption, the shearing (i.e. tangential) stress between adjacent layers of fluid of infinitesimally small thickness is taken to be proportional to the rate of shear in the direction perpendicular

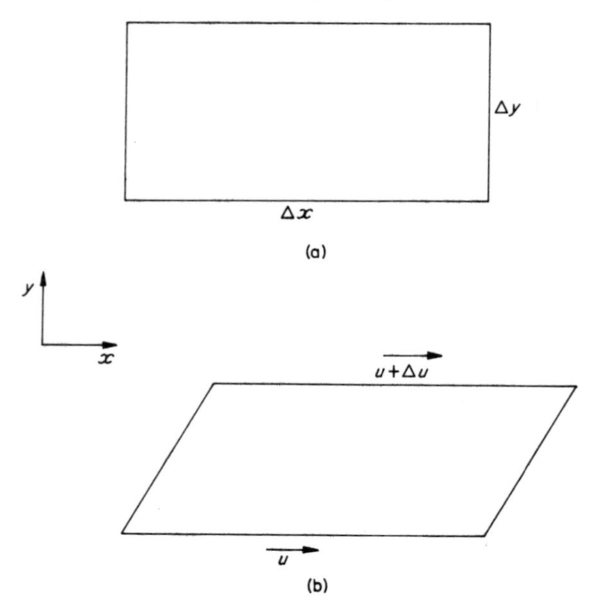

Fig. 2.1 Particle of viscous fluid set in motion

to the motion. Thus, consider a fluid particle, as shown in Fig. 2.1(a), whose sides are Δx and Δy. If it is set in motion in the x direction with a velocity u on its lower surface and $u + \Delta u$ on the upper surface, its shape will change to that shown in Fig. 2.1(b). The rate of shear in the direction normal to the motion is then $\Delta u/\Delta y$, so that the limiting value of the rate of shear for an infinitesimally small particle is $\partial u/\partial y$. Use of the second assumption for the shearing stress τ at any point leads to the equation

$$\tau = \mu \frac{\partial u}{\partial y}$$

where μ is a coefficient of proportionality, usually termed the absolute viscosity of the fluid. A fluid which obeys such a relationship is known as a Newtonian fluid. ECM electrolytes, water, air, and gases can all be regarded as Newtonian.

Thus

$$\mu = \tau \bigg/ \frac{\partial u}{\partial y} \qquad (2.1)$$

That is, the viscosity is the ratio of shear stress to transverse velocity gradient.

(a) Kinematic viscosity

It is often convenient to deal with the kinematic viscosity, ν, obtained by dividing μ by the density, ρ_e, of the fluid:

$$\nu = \frac{\mu}{\rho_e} \tag{2.2}$$

(b) Specific viscosity

This is defined as ratio of the absolute viscosity of the fluid to that of water at 20°C.

The viscosity of water at 20°C is almost 1 cP so that the viscosity in centipoise is numerically equal to the specific viscosity. For liquids, μ is nearly independent of pressure and decreases rapidly as the temperature increases; ν has the same dependence. The liquid electrolytes used in ECM, of course, obey these conditions. A particular requirement for them is that their viscosity should be low, so that they can be pumped at high rates down the inter-electrode gap without the need for excessively high pump pressures. For example, for 10% (w/w) NaCl solution in water, μ = 1·23 cP at 18°C; at 40°C, μ = 0·78 cP. Since ρ_e = 1·07 g/cm^3 at 18°C and 1·06 g/cm^3 at 40°C, the corresponding ν values are 1·15 mm^2/s and 0·735 mm^2/s respectively. For gases, μ increases with temperature, although it is again independent of pressure; ν increases rapidly with temperature since the density decreases rapidly with increasing temperature.

(c) Compressibility

Compressibility gives a measure of the change of volume of a fluid subjected to normal pressures or tensions. It can be defined thus: compressibility = percentage change in volume for a given pressure change.

For water, a pressure increase of 101 kN/m^2 causes a relative change in volume of about 0·005%. This low value is typical of most liquids, so that they may be regarded as incompressible. Air, however, at NTP is about 20 000 times more compressible than water. This is typical for most gases. In the work which follows we shall assume that the fluid is incompressible, and consider implicitly that it is an electrolyte.

(d) Reynolds number

The dimensionless quantity

$$\rho_e \frac{Ul}{\mu} = \frac{Ul}{\nu} = \mathrm{Re} \tag{2.3}$$

where l and U are a characteristic length and velocity, respectively, for the flow path, is termed the Reynolds number, Re.

(e) Hydraulic diameter

In ECM, the electrolyte often flows through channels of non-circular cross-section. It is convenient then to define a 'hydraulic diameter', d_h,

$$d_h = \frac{4A}{C} \tag{2.4}$$

where A is the cross-section area and C is the wetted perimeter.

For a rectangular flow channel, formed by plane, parallel electrodes, where the gap h is much smaller than the electrode width, b

$$d_h = 4bh/2(b+h) \simeq 2h \tag{2.5}$$

2.2 Navier-Stokes equations

The flow of a viscous fluid like an electrolyte can be specified by the three orthogonal components (u, v, w) of its velocity U, the pressure $p(x, y, z, t)$, and density $\rho_e(x, y, z, t)$, where x, y, z are the usual orthogonal coordinates of position and t is the time. Five equations are available for the determination of u, v, w, p, and ρ_e. They are (i) the three equations of motion for the conservation of momentum, (ii) the continuity equation for the conservation of mass, and (iii) the thermodynamic equation of state.

The equations of motion are obtained from Newton's second law. In the application of this law to a fluid in motion, two types of forces must be considered: (a) body (e.g. gravitational) forces, which act throughout the mass of the body, and (b) pressure and friction forces, which act on its boundary. The equations of motion can then be expressed in the form

$$\rho_e \frac{dU}{dt} = F + P \qquad (2.6)$$

where F is the gravitational force per unit volume ($\rho_e g$, g being the acceleration due to gravity) and P is the boundary force. As usual, F and P can be written in component form:

$$F = iF_x + jF_y + kF_z \qquad (2.7)$$

and

$$P = iP_x + jP_y + kP_z \qquad (2.8)$$

where i, j, and k are unit vectors along the x, y, and z axes. It should be noted that these are the only types of forces considered in this analysis. Little is yet known about the effect on the fluid motion of other forces that may arise in ECM, for instance, those of an electromagnetic nature.

The thermodynamic equation of state can be shown to have the form

$$\mathrm{F}(\rho, p, T) = 0 \qquad (2.9)$$

For instance, for a perfect gas, the usual equation of state is

$$p = \rho_g RT \qquad (2.10)$$

where ρ_g is the gas density, R is the gas constant, and T the temperature.

Discussion of these equations is available elsewhere [1]. Since we are mainly concerned with liquid electrolytes, it is sufficient for our purposes to write down the fundamental equations of motion for incompressible flow:

$$\rho_e\left(\frac{\partial u}{\partial t} + u\frac{\partial u}{\partial x} + v\frac{\partial u}{\partial y} + w\frac{\partial u}{\partial z}\right) = F_x - \frac{\partial p}{\partial x} + \mu\left(\frac{\partial^2 u}{\partial x^2} + \frac{\partial^2 u}{\partial y^2} + \frac{\partial^2 u}{\partial z^2}\right) \qquad (2.11)$$

$$\rho_e\left(\frac{\partial v}{\partial t} + u\frac{\partial v}{\partial x} + v\frac{\partial v}{\partial y} + w\frac{\partial v}{\partial z}\right) = F_y - \frac{\partial p}{\partial y} + \mu\left(\frac{\partial^2 v}{\partial x^2} + \frac{\partial^2 v}{\partial y^2} + \frac{\partial^2 v}{\partial z^2}\right) \qquad (2.12)$$

$$\rho_e\left(\frac{\partial w}{\partial t} + u\frac{\partial w}{\partial x} + v\frac{\partial w}{\partial y} + w\frac{\partial w}{\partial z}\right) = F_z - \frac{\partial p}{\partial z} + \mu\left(\frac{\partial^2 w}{\partial x^2} + \frac{\partial^2 w}{\partial y^2} + \frac{\partial^2 w}{\partial z^2}\right) \qquad (2.13)$$

$$\frac{\partial u}{\partial x} + \frac{\partial v}{\partial y} + \frac{\partial w}{\partial z} = 0 \qquad (2.14)$$

The first three equations are the Navier–Stokes equations. The fourth is the equation of continuity.

2.3 Laminar flow

In ECM the flow is likely to be either *laminar* or *turbulent*, although the latter will be shown later to be more common. In laminar flow, agitation of the fluid particles is of a molecular nature, and these particles are constrained to motion in parallel paths, or layers, by the action of viscosity. The shearing stress between adjacent moving layers is determined by the viscosity, and is defined by Equation (2.1). Some characteristic behaviour can be obtained from an examination of laminar flow down a straight channel.

2.3.1 Laminar flow down a straight channel

Consider two-dimensional, steady, laminar flow down a straight channel with parallel flat walls, spaced distance h apart. Since the flow is steady, all time derivatives of the velocity are zero. If the flow is also assumed to be parallel in the x direction so that there is only one velocity component, the equation of continuity (2.14) gives that

$$\frac{\partial u}{\partial x} = 0$$

Then

$$u = u(y) \quad \text{and} \quad \frac{\partial^2 u}{\partial x^2} = 0$$

$$v = 0$$

$$w = 0$$

From the equations of motion (2.12) and (2.13) for the y and z directions,

$$\frac{\partial p}{\partial y} = \frac{\partial p}{\partial z} = 0; \quad \frac{\partial p}{\partial x} = \text{constant.}$$

The Navier–Stokes equation (2.11) for the x direction then becomes

$$\frac{\mathrm{d}p}{\mathrm{d}x} = \mu \frac{\mathrm{d}^2 u}{\mathrm{d}y^2} \tag{2.15}$$

The solution to Equation (2.15) is

$$u = \frac{1}{2\mu}\left(-\frac{dp}{dx}\right)\left(\frac{h^2}{4} - y^2\right) \tag{2.16}$$

the boundary conditions being $u = 0$ at $y = \pm h/2$. The maximum velocity, u_{max}, is given by

$$u_{max} = \frac{h^2}{8\mu}\left(-\frac{dp}{dx}\right) \tag{2.17a}$$

whilst the average velocity $\bar{u}$ is

$$\bar{u} = \tfrac{2}{3}\mu_{max}$$

$$= \frac{h^2}{12\mu}\left(-\frac{dp}{dx}\right) \tag{2.17b}$$

From Equation (2.16), the velocity distribution is deduced to be parabolic. This type of velocity profile, which is shown in Fig. 2.2, is a characteristic of laminar flow.

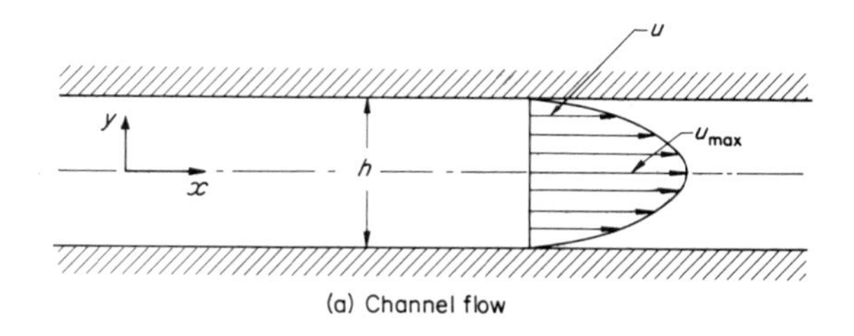

(a) Channel flow

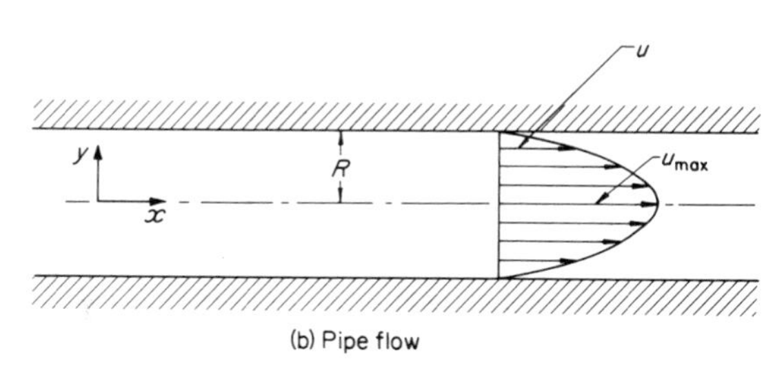

(b) Pipe flow

Fig. 2.2 Parabolic velocity distribution for laminar flow

2.3.2 Hagen–Poiseuille flow

The above analysis on flow down a channel can be extended to cover steady laminar flow due to a pressure drop along a straight pipe of circular cross-section. This is usually termed Hagen–Poiseuille flow. The velocity distribution may be obtained by putting the Navier–Stokes equations into cylindrical coordinates. The equations then reduce to one equation for the x direction (taken to be coincident with the axis of the pipe). Since the tangential and radial velocity components are zero, the velocity component in the x direction depends only on y, and the pressure gradient $\mathrm{d}p/\mathrm{d}x$ is constant. The equation can be written:

$$\frac{\mathrm{d}p}{\mathrm{d}x} = \mu\left(\frac{\mathrm{d}^2 u}{\mathrm{d}y^2} + \frac{1}{y}\frac{\mathrm{d}u}{\mathrm{d}y}\right)$$

where $u = 0$ at $y = R$, R being the radius of the pipe. The solution is

$$u(y) = -\frac{1}{4\mu}\frac{\mathrm{d}p}{\mathrm{d}x}(R^2 - y^2)$$

Here, the constant pressure gradient, $-(\mathrm{d}p/\mathrm{d}x) = (p_1 - p_2)/L$, is assumed to be known, L being the pipe length.

As before, the maximum velocity $u_{\max}$ can be calculated:

$$u_{\max} = \frac{R^2}{4\mu}\left(-\frac{\mathrm{d}p}{\mathrm{d}x}\right) \tag{2.18}$$

whilst the mean velocity $\bar{u}$ is

$$\begin{aligned}\bar{u} &= \tfrac{1}{2}u_{\max}\\ &= \frac{R^2}{8\mu}\left(-\frac{\mathrm{d}p}{\mathrm{d}x}\right)\end{aligned} \tag{2.19}$$

This equation, and that above for flow down a channel, Equation (2.17b), are used to calculate the pressure difference required to overcome viscous forces in laminar flow.

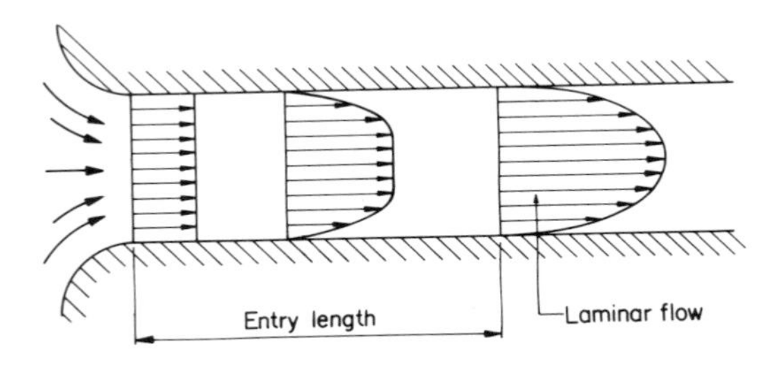

Fig. 2.3 Establishment of fully developed laminar flow

Note that in Sections 2.3.1 and 2.3.2, the average velocity can be expressed in terms of the volumetric flow-rate Q, the flow quantity usually measured in ECM. For instance, in pipe flow, $Q = \pi R^2 \bar{u}$.

In the above analyses, the laminar flow has been assumed to be fully developed. This condition is only achieved at some distance from the entrance to the channel. In Fig. 2.3, it is seen that the velocity distribution in a pipe can change from an almost uniform profile at the smoothly curved inlet to a fully developed profile further downstream. The transition is caused by an increase in viscous effects along the channel. The entry length, x_E, required for fully developed flow was established theoretically by Boussinesq as

$$x_E = 0{\cdot}03 \, d \, \mathrm{Re} \tag{2.20}$$

where d is the pipe diameter and Re the Reynolds number ($= \bar{u}d/\nu$). Thus, for Re = 1000, x_E is about 30 pipe diameters. [No equivalent formula to (2.20) exists for turbulent flow, although experiments have shown that x_E is often about 50 to 100 diameters.]

From the discussion of 'internal' flow in channels and pipes, we move to considerations of flow around objects, or 'external' flow. Although the latter flow can be described by the same equations of motion used for internal flow, its characteristics are quite different. A typical example of external flow is flow over a flat plate. Closely associated with this kind of flow is the phenomenon of boundary layers.

2.4 Boundary layers in laminar flow

2.4.1 Thickness of laminar boundary layer

For an electrolyte, flowing over a rigid body (for example, a flat plate), the 'no slip' condition applies at the surface of the body. That is, the fluid velocity at that surface is zero. Experiments show that the fluid velocity increases from zero at the boundary to its mainstream value over a thin layer, termed the 'boundary layer'. Over the boundary layer, the fluid velocity gradient is high, and viscous effects have as much influence as inertia effects. Outside the layer, the velocity gradients and the viscous shear are small. The flow is then influenced by inertia, pressure gradient, and body forces. The importance of the boundary layer in ECM will become apparent in later chapters. At this stage it is useful to obtain an expression for its magnitude, δ_0.

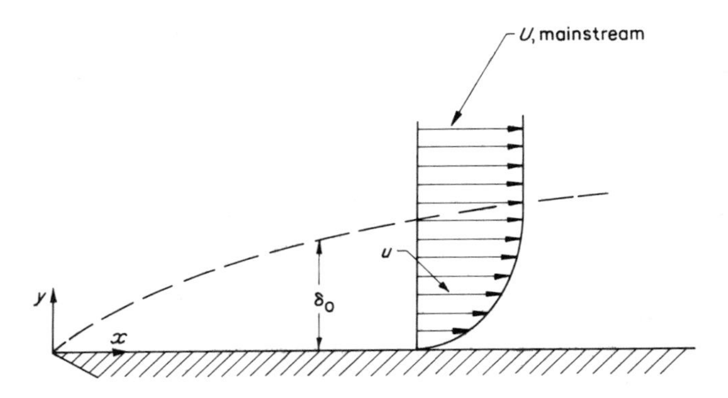

Fig. 2.4 Velocity distribution in the boundary layer

Let us examine the Navier–Stokes equations for the boundary layer for two-dimensional, steady incompressible flow over a flat plate (Fig. 2.4). Since body forces can be ignored, then, from Equations (2.11) and (2.12):

$$\rho_e\left(u\frac{\partial u}{\partial x}+v\frac{\partial u}{\partial y}\right)=-\frac{\partial p}{\partial x}+\mu\left(\frac{\partial^2 u}{\partial x^2}+\frac{\partial^2 u}{\partial y^2}\right) \tag{2.21}$$

$$\rho_e\left(u\frac{\partial v}{\partial x}+v\frac{\partial v}{\partial y}\right)=-\frac{\partial p}{\partial y}+\mu\left(\frac{\partial^2 v}{\partial x^2}+\frac{\partial^2 v}{\partial y^2}\right) \tag{2.22}$$

The continuity equation (2.14) reduces to

$$\frac{\partial u}{\partial x}+\frac{\partial v}{\partial y}=0 \tag{2.23}$$

Let us now carry out a rough order of magnitude study of the terms in the above equations. First it is noted that, since δ_0/x is usually very small, the boundary layer is thin, x being a distance measured from the leading edge of the plate. A time scale is now chosen so that U, the mainstream velocity, is the same order of magnitude as x. Suppose, too, that x is of order unity. We write

$$U \simeq x = 0(1)$$

Now magnitudes of distances in the y direction within the boundary layer are much less than x. These are denoted by $0(\delta_0)$. That is

$$0(\delta_0) \ll 0(1)$$

The extreme value of the difference of the velocity u in the y direction is also $0(1)$, since this velocity goes from zero to U. This condition also holds for the second difference of $u(y)$. Since the extreme difference of y is $0(\delta_0)$,

$$\frac{\partial u}{\partial y} = 0\left(\frac{1}{\delta_0}\right)$$

$$\frac{\partial^2 u}{\partial y^2} = 0\left(\frac{1}{\delta_0^2}\right)$$

Next, u can change from almost U at $x = 0$ to almost zero at x. Thus the extreme magnitude of $\partial u/\partial x$ is unity. Then

$$\frac{\partial u}{\partial x} = 0(1)$$

Similarly,

$$\frac{\partial^2 u}{\partial x^2} = 0(1)$$

The continuity equation (2.14) shows that since $\partial u/\partial x$ is $0(1)$, so must be $\partial v/\partial y$. Since changes in y are $0(\delta_0)$, changes in $v(y)$ are also $0(\delta_0)$. Finally, since $v = 0$ for $y = 0$, $v = 0(\delta_0)$. We may now write

$$v = 0(\delta_0)$$

$$\frac{\partial v}{\partial y} = 0(1)$$

$$\frac{\partial^2 v}{\partial y^2} = 0\left(\frac{1}{\delta_0}\right)$$

$$\frac{\partial v}{\partial x} = 0(\delta_0)$$

$$\frac{\partial^2 v}{\partial x^2} = 0(\delta_0)$$

In the original equation (2.21) we then have

$$u\frac{\partial u}{\partial x} + v\frac{\partial u}{\partial y} = -\frac{1}{\rho_e}\frac{\partial p}{\partial x} + \frac{\mu}{\rho_e}\left(\frac{\partial^2 u}{\partial x^2} + \frac{\partial^2 u}{\partial y^2}\right)$$

$$[0(1)]\,[0(1)] + [0(\delta_0)][0(1/\delta_0)] = -\frac{1}{\rho_e}\frac{\partial p}{\partial x} + \frac{\mu}{\rho_e}([0(1)] + [0(1/\delta_0^2)]) \qquad (2.24)$$

Since terms of $0(1)$ are much less than terms of $0(1/\delta_0^2)$, $\partial^2 u/\partial x^2$ can be neglected in comparison with $\partial^2 u/\partial y^2$.

But the expression $(\partial^2 u/\partial x^2 + \partial^2 u/\partial y^2)$ is derived from friction effects which are considerable in the boundary layer. It must, therefore, have an order of magnitude similar to those of the other expressions in Equation (2.24). Those expressions have an order of magnitude of unity. Thus

$$\nu[0(1/\delta_0^2)] = 0(1)$$

where $\mu/\rho_e = \nu$, the kinematic viscosity. Accordingly,

$$\nu = 0(\delta_0^2)$$

Note, too, that a Reynolds number Re of the form Ux/ν is of order $0(1/\delta_0^2)$. That is, δ_0 is related to Re by

$$\delta_0 = 0(\mathrm{Re}^{-1/2})$$

and since x is of order of magnitude unity,

$$\frac{\delta_0}{x} = 0(\mathrm{Re}^{-1/2})$$

Insertion of the experimentally and theoretically derived, numerical coefficient gives the result:

$$\delta_0 = 5\left(\frac{\nu x}{U}\right)^{1/2} \tag{2.25}$$

This equation describes the distance from the wall at which the boundary layer has sensibly merged with the mainstream.

2.4.2 Prandtl's boundary layer equations

Consider the second of the Navier–Stokes equations (2.12) and the corresponding order of magnitude:

$$u\frac{\partial v}{\partial x} + v\frac{\partial v}{\partial y} = -\frac{1}{\rho_e}\frac{\partial p}{\partial y} + \nu\left(\frac{\partial^2 v}{\partial x^2} + \frac{\partial^2 v}{\partial y^2}\right)$$

$$[0(1)][0(\delta_0)] + [0(\delta_0)][0(1)] = -\frac{1}{\rho_e}\frac{\partial p}{\partial y} + [0(\delta_0^2)][0(\delta_0) + 0(1/\delta_0)] \tag{2.26}$$

Note that $\partial^2 v/\partial x^2$ is negligible compared with $\partial^2 v/\partial y^2$. Since the term $[-(1/\rho_e)(\partial p/\partial y)]$ is significant, it will be $0(\delta_0)$. Now ρ_e can be taken to be $0(1)$ or less, thus $\partial p/\partial y$ is $0(\delta_0)$. It is also clear that the velocity v which is $O(\delta_0)$ is negligible compared with u, which is $O(1)$. We thus can neglect Equation (2.26) on the basis of an effectively constant pressure through the boundary layer. Then we consider only the previous equation (2.24) with $\partial^2 u/\partial x^2$ deleted. That is

$$u\frac{\partial u}{\partial x} + v\frac{\partial u}{\partial y} = -\frac{1}{\rho_e}\frac{dp}{dx} + \nu\frac{\partial^2 u}{\partial y^2} \tag{2.27}$$

$$\frac{\partial u}{\partial x} + \frac{\partial v}{\partial y} = 0 \tag{2.28}$$

Equations (2.27) and (2.28) are known as Prandtl's boundary layer equations for steady flow.

2.4.3 Bernoulli's equation

Near the outer edge of the boundary layer, $u = U$. Over this region there is no large velocity gradient and the viscous terms in Equation (2.27) can be ignored. We obtain

$$U\frac{\partial U}{\partial x} = -\frac{1}{\rho_e}\frac{\partial p}{\partial x} \tag{2.29}$$

Integration gives

$$p + \tfrac{1}{2}\rho_e U^2 = \text{constant} \tag{2.30}$$

This is Bernoulli's equation, from which the pressure required to overcome inertia can be calculated.

2.5 Transition from laminar to turbulent flow

From the Hagen–Poiseuille equation, we have seen that the velocity distribution for laminar flow is parabolic. The usual criterion for the existence of laminar flow down a pipe of diameter d is that

$$\text{Re} = \frac{\bar{u}d}{\nu} < 2300 \tag{2.31}$$

where $\bar{u}$ is the mean velocity. Above Re = 2300 the mean velocity distribution is likely to be more uniform and the flow is then found to be *turbulent.* Figure 2.5 demonstrates the difference in velocity profiles between laminar and turbulent flow.

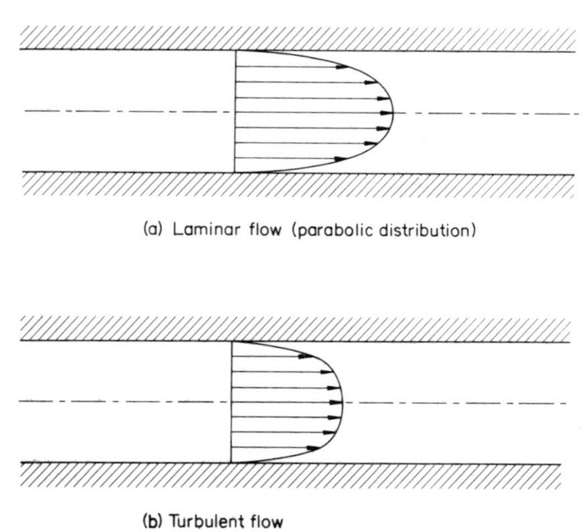

Fig. 2.5 Velocity distributions in a pipe

2.6 Turbulent flow

In turbulent flow the fluid particles have a random irregular motion, and at any point in the fluid, their velocity and pressure vary in both time and direction. In any analysis, these physical features of the motion are usually separated into a 'mean component' and a 'fluctuation component', for instance, any mean velocity component and its velocity fluctuation component. This fluctuation affects the mean motion such that there appears an apparent increase in the viscosity of the main flow. The mean velocity components can be shown to satisfy the same equations as those of laminar flow except that the laminar stresses are increased by additional stresses due to turbulent fluctuations; they are known as 'apparent stresses of turbulent flow', or Reynolds stresses. In most flows, the apparent stresses are greater than the viscous components and so the latter may often be omitted.

The no-slip condition means that the mean velocity components must be zero at the wall. The turbulent components also vanish there, and are very small in its vicinity. Accordingly, all components of apparent stress also vanish at the walls and only the viscous stresses of laminar flow are active. It follows that, next to the wall, the viscous stresses predominate over the apparent stresses, and that there should exist a thin layer whose motion is similar to a laminar one. This is the 'laminar sub-layer'. Since the velocities within it are so small, the viscous forces are dominant and there is no turbulence in it. The laminar sub-layer merges into another layer in which the velocity fluctuations are so great that turbulent shearing

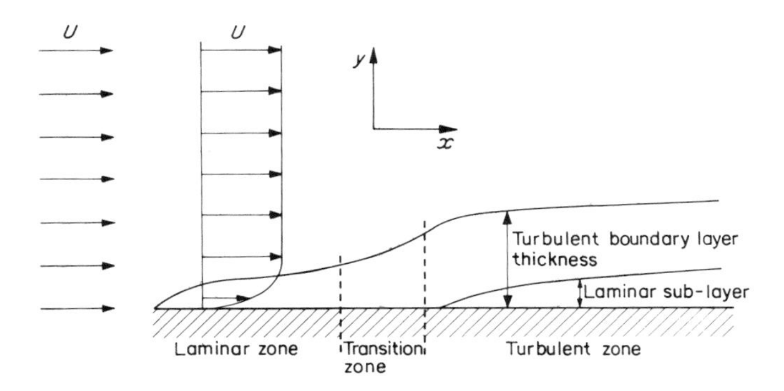

Fig. 2.6 Development of turbulent boundary layer

stresses, comparable to viscous stresses, arise. At larger distances from the wall, the turbulent stresses far exceed the viscous stresses. The turbulent boundary layer is comprised of these distances. The various layers are illustrated in Fig. 2.6 which shows the development of a turbulent boundary layer along a flat plate. At the leading edge, a laminar boundary layer starts to grow. Downstream, a transition region is reached where the flow changes from laminar to turbulent. Beyond this region, the turbulence is fully developed with the laminar sub-layer formed along the wall; and outside the latter layer is the turbulent boundary layer proper. In the next few sections equations will be deduced from which the sizes of the laminar sub-layer and the turbulent boundary layer can be estimated. But first an expression is derived for the pressure gradient down a pipe.

2.6.1 *Pressure gradient in turbulent flow*

We concern ourselves with fully developed turbulent flow down a smooth pipe of diameter d. The relationship between the pressure gradient and the mean velocity $\overline{u}$ is usually determined from the laws of friction. To obtain this relationship, it is useful to define a dimensionless coefficient of resistance f:

$$\left(-\frac{\mathrm{d}p}{\mathrm{d}x}\right) = \frac{f\rho_{\mathrm{e}}}{2d}\,\overline{u}^2 \qquad (2.32)$$

It is relevant also to introduce here the empirical formula established by Blasius, which relates f to Reynolds number for smooth pipes of circular cross-section:

$$f = 0{\cdot}3164\left(\frac{\bar{u}d}{\nu}\right)^{-1/4} \tag{2.33}$$

where $\mathrm{Re} = \bar{u}d/\nu$, and $\bar{u} = Q/\pi R^2$ where Q is the volume flow-rate through the pipe. This formula is valid for Reynolds numbers ranging from 3 000 to 100 000.

The pressure drop in turbulent flow is deduced to be:

$$\left(-\frac{\mathrm{d}p}{\mathrm{d}x}\right) = 0{\cdot}3164\,\frac{\rho_e \bar{u}^2}{2d}\left(\frac{\bar{u}d}{\nu}\right)^{-1/4} \tag{2.34}$$

This equation corresponds to Equations (2.17b) and (2.19). From it, the pressure drop required to overcome viscous forces in turbulent flow can be calculated.

2.6.2 *Velocity distribution laws*

Experimental investigations have demonstrated that for turbulent flow through a smooth pipe, the velocity ratio u/u_{max} can be related to the distance ratio y/R by an empirical expression of the form

$$\frac{u}{u_{max}} = \left(\frac{y}{R}\right)^{1/n}$$

where u is the local velocity at distance y, and R is the pipe radius. The exponent n varies with Reynolds number: calculations of the ratio u/u_{max} have shown that $n = 6$ and $n = 7$ for Re = 4 000 and Re = 100 000, respectively. Table 2.1 shows the variation in n as a function of the ratio of mean velocity to maximum velocity.

Table 2.1 Variation in n as a function of the ratio of mean to maximum velocity in a pipe (after Schlichting [1])

n	6	7	8	9	10
$\bar{u}/u_{max}$	0·791	0·817	0·837	0·852	0·865

[illegible] the Blasius formula (2.33), [illegible]d which holds for short [illegible] friction velocity, u_f, is defined:

$$[illegible]\Big)^{1/2} \qquad (2.35)$$

[illegible]e wall of the pipe. Earlier in this [illegible]es are negligible in comparison with [illegible]e wall. The condition of equilibrium betw[illegible]earing stress τ_0 on the circumference and the forc[illegible] due to the pressure gradient $(-\mathrm{d}p/\mathrm{d}x)$ then is

$$\tau_0 = \left(-\frac{\mathrm{d}p}{\mathrm{d}x}\right)\frac{R}{2} \qquad (2.36)$$

Substitution for $(-\mathrm{d}p/\mathrm{d}x)$ from Equation (2.34) into Equation (2.36) gives

$$\tau_0 = 0{\cdot}0395\, \rho_e \bar{u}^{7/4}\, \nu^{1/4}\, d^{-1/4} \qquad (2.37)$$

Use of Equation (2.35) then leads to

$$\left(\frac{\bar{u}}{u_f}\right) = 6{\cdot}99\left(\frac{u_f R}{\nu}\right)^{1/7} \qquad (2.38)$$

The results shown in Table 2.1 now allow the replacement of $\bar{u}$ by u_{max}: $\bar{u} = 0{\cdot}8u_{max}$ for $n = 7$. We have

$$\frac{u_{max}}{u_f} = 8{\cdot}74\left(\frac{u_f R}{\nu}\right)^{1/7} \qquad (2.39)$$

This equation is usually assumed to hold for any distance y (and not only $y = R$) and for any velocity $u(y)$. That is,

$$\frac{u(y)}{u_f} = 8{\cdot}74\left(\frac{u_f y}{\nu}\right)^{1/7} \qquad (2.40)$$

Equations (2.39) and (2.40) are based on the Blasius friction factor relationship (2.33) which is valid for Reynolds numbers up to about 100 000 and for values of $(u_f y/\nu)$ ranging from about 500 to 2000. Above these values, viscous effects no longer predominate over turbulent stresses and an alternative relationship has to be found. Schlichting [1] quotes the usefulness of these conditions:

Laminar sub-layer: $0 < u_f y/\nu < 4$ to 5 – only laminar friction exists.
Buffer region: 4 to $5 < u_f y/\nu < 30$ to 70 – laminar and turbulent friction are of the same order of magnitude.
Turbulent region: $u_f y/\nu > 30$ to 70 – turbulent friction is much greater than laminar friction.

Equations (2.39) and (2.40) indicate that the velocity distribution can be written as a power law. The experimental evidence, that the exponent in those equations decreases as the Reynolds number is increased, suggests that a universal law may exist which is valid for all Reynolds numbers. Such a law, in logarithmic form, has been developed which is applicable, in addition, to other channel flows and to two-dimensional boundary layers. A generally accepted version of the law is

$$\frac{U - u(y)}{u_f} = C_1 \ln \frac{y}{y_0} \tag{2.41}$$

where U now represents the maximum (centre-line) velocity for pipe flow and the mainstream velocity for boundary layer flow. C_1 is an experimental constant, usually of order $-2{\cdot}5$, and y_0 is some suitable reference length. (Note: in most ECM work, U and $\bar{u}$ are regarded as equivalent; both are usually calculated from the volumetric flow-rate at the gap inlet, divided by the cross-sectional area of the channel at inlet.)

2.6.3 Thickness of laminar sub-layer

The universal formula allows an expression for the thickness δ_l of the laminar sub-layer to be obtained. (This quantity must be known for the calculations of concentration overpotential in the next chapter.) From the condition for the existence of laminar friction in Section 2.6.2,

$$\delta_l \simeq \frac{5\nu}{u_f} \tag{2.42}$$

If the velocity $u(y)$ at the edge of the laminar sub-layer is assumed to be small compared with the mainstream velocity U, the friction velocity is given from Equation (2.41) as

$$\frac{U}{u_f} = C_1 \ln \frac{\delta_l}{y_0} \tag{2.43}$$

As always, difficulty arises in the selection of the reference length y_0. In ECM and for flow down a rectangular channel, a convenient choice for y_0 is the inter-electrode gap width h. From the discussion in Section 2.1, $y_0 = 2h$. By means of Equations (2.42) and (2.43) we deduce that

$$\delta_l = \frac{5\nu C_1}{U} \ln \frac{\delta_l}{2h} \tag{2.44a}$$

(Note: although C_1 is usually about -2.5, the magnitude of δ_l is not greatly affected for C_1 varying between 1 and 10.)

An alternative treatment based on the same reasoning is available [3] which gives the result

$$\frac{\delta_l}{R} = 124 \text{ Re}^{-7/8} \tag{2.44b}$$

where R is the channel radius and $\text{Re} = Ud_h/\nu$, is the Reynolds number.

2.6.4 *Thickness of the turbulent boundary layer*

The exact thickness, δ_t of the boundary layer cannot be found since the division between the zones of negligible viscosity and the boundary layer is a gradual one. However, δ_t is often defined as the distance at which the velocity reaches an arbitrary percentage fraction of the mainstream velocity (e.g. 99%). This condition is indicated in Fig. 2.7. Two other 'thicknesses' can usefully be defined.

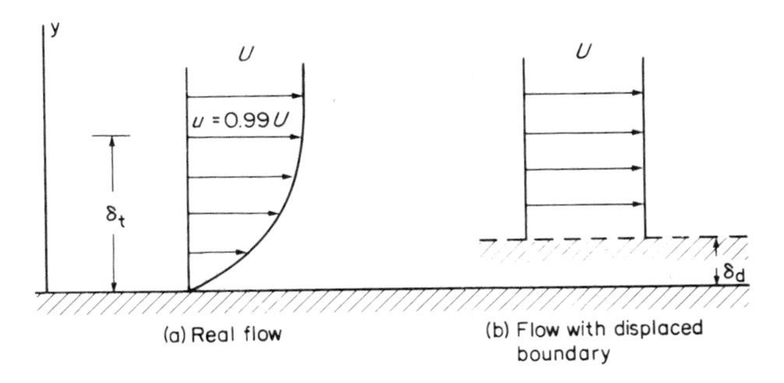

Fig. 2.7 Thickness of turbulent boundary layer

One is the displacement thickness, δ_d, the distance by which the boundary would be displaced to maintain the same mass flow at any section, the entire flow being assumed to be frictionless. From Fig. 2.7, by the continuity of mass, δ_d is given by

$$\rho_e U \delta_d = \rho_e \int_0^{\delta_t{}^+} (U - u)\, dy$$

i.e.

$$\delta_d = \int_0^{\delta_t{}^+} \left(1 - \frac{u}{U}\right) dy \qquad (2.45)$$

where $\delta_t + \geqslant \delta_t$. U, as usual, is the mainstream velocity and u the velocity within the boundary layer.

A reduction in the momentum flux rate due to the reduction in velocity within the boundary layer gives rise to the 'momentum thickness' θ, defined as

$$\rho_e \theta U^2 = \rho_e \int_0^{\delta_t{}^+} (Uu - u^2)\, dy$$

or

$$\theta = \int_0^{\delta_t{}^+} \frac{u}{U}\left(1 - \frac{u}{U}\right) dy \qquad (2.46)$$

To illustrate the physical significance of the momentum thickness, a relationship is derived between the shear stress at the boundary and the velocity. From Prandtl's boundary layer equations (2.27) and (2.28), and in the absence of pressure gradient, integration gives

$$\int_{y=0}^{y=\delta_t{}^+} \left(u\frac{\partial u}{\partial x} + v\frac{\partial u}{\partial y}\right) dy = \nu \int_0^{\delta_t{}^+} \frac{\partial^2 u}{\partial y^2}\, dy$$

From Equation (2.1)

$$\mu \int_0^{\delta_t{}^+} \frac{\partial^2 u}{\partial y^2} = \int_0^{\delta_t{}^+} \frac{\partial \tau}{\partial y}\, dy = [0 - \tau_0] = -\tau_0$$

Now

$$\int_0^{\delta_t{}^+} v\frac{\partial u}{\partial y}\, dy = \int_0^{\delta_t{}^+} \frac{\partial}{\partial y}(uv)\, dy - \int_0^{\delta_t{}^+} u\frac{\partial v}{\partial y}\, dy$$

$$= Uv_{\delta_t{}^+} + \int_0^{\delta_t{}^+} u\frac{\partial u}{\partial x}\, dy$$

by use of the continuity equation and where

$$v_{\delta_t^+} = -\int_0^{\delta_t^+} \frac{\partial u}{\partial x}\,dy$$

Thus

$$\int_0^{\delta_t^+} u\frac{\partial u}{\partial x}\,dy - U\int_0^{\delta_t^+} \frac{\partial u}{\partial x}\,dy + \int_0^{\delta_t^+} u\frac{\partial u}{\partial x}\,dy = -\frac{\tau_0}{\rho_e}$$

If U is constant, this equation becomes

$$\frac{\tau_0}{\rho_e} = \frac{\partial}{\partial x}\int_0^{\delta_t^+} u(U-u)\,dy \tag{2.47}$$

That is,

$$\tau_0 = \rho_e U^2 \frac{\partial \theta}{\partial x}$$

$$= \rho_e U^2 \frac{d\theta}{d\delta_t}\frac{d\delta_t}{dx}$$

Now, from Equations (2.35) and (2.40), and assuming that those analyses for pipe flow can again be applied to flow over a flat plate, we deduce that

$$\tau_0 = 0{\cdot}0225\,\rho_e U^2 \left(\frac{\nu}{U\delta_t}\right)^{1/4} \tag{2.48}$$

The power law derived above can be adapted further. From Equation (2.40),

$$\frac{U}{u_f} = 8{\cdot}74\left(\frac{u_f \delta_t}{\nu}\right)^{1/7}$$

and

$$\frac{u(y)}{u_f} = 8{\cdot}74\left(\frac{u_f y}{\nu}\right)^{1/7}$$

Combining these equations, we obtain

$$\frac{u(y)}{U} = \left(\frac{y}{\delta_t}\right)^{1/7} \tag{2.49}$$

Substitution for $u(y)$ and for τ_0 in Equation (2.48) gives

$$0{\cdot}0225\ U^2\left(\frac{\nu}{U\delta_t}\right)^{1/4} = \frac{d}{dx}\int_0^{\delta_t^+} U^2\left[\left(\frac{y}{\delta_t}\right)^{1/7} - \left(\frac{y}{\delta_t}\right)^{2/7}\right] dy$$

Integration yields

$$0{\cdot}0225\left(\frac{\nu}{U\delta_t}\right)^{1/4} = +\frac{7}{72}\frac{d\delta_t}{dx}$$

Separation of variables and integration again lead to

$$\left(\frac{\nu}{U}\right) x^{1/4} = 3{\cdot}45\ \delta_t^{5/4} + C \qquad (2.50)$$

where C is the constant of integration. This constant is difficult to find, since the turbulent boundary layer starts at the transition zone, downstream from the leading edge; location of that zone is difficult, and moreover, the layer has a finite thickness there. However, it is recognised that if the boundary layer is assumed to start at the edge of the plate, acceptable results can be obtained for points beyond the transition region. If this procedure is followed, and $\delta_t = 0$ at $x = 0$, then $C = 0$. Equation (2.50) becomes

$$\delta_t = 0{\cdot}376\ x^{4/5}\left(\frac{\nu}{U}\right)^{1/5} \qquad (2.51)$$

2.7 Admissible surface roughness

In the above analyses, and in ECM calculations based upon them, the plate and pipe surfaces have been implicitly assumed to be hydraulically smooth. Some surface roughness is usually thought to be tolerable, so that the condition for hydraulic smoothness can still be maintained, provided the height of the roughness irregularities is below a critical amount. This amount also appears to depend on the type of flow.

For turbulent flow along a pipe and a flat plate, one recommended relationship is

$$\frac{U\epsilon_{adm}}{\nu} = 10^2 \qquad (2.52)$$

where U is the mainstream velocity, ϵ_{adm} is the admissible height of the surface projections, and ν is the kinematic viscosity. The following corresponding relationship for laminar flow has also been suggested:

$$\frac{u_f \epsilon_{adm}}{\nu} = 15 \tag{2.53}$$

where u_f is the friction velocity, Since

$$u_f = \left(\frac{\tau_0}{\rho_e}\right)^{1/2} \tag{2.35}$$

It can also be shown that the shearing stress (τ_0/ρ_e) is given by

$$\frac{\tau_0}{\rho_e} = 0{\cdot}332\ U^2\left(\frac{\nu}{Ux}\right)^{1/2} \tag{2.54}$$

[cf. Equation (2.48) for turbulent flow]. Hence, ϵ_{adm} can be expressed in terms of U, ν, ρ_e, and x.

The usefulness of these hypotheses remains to be tested in ECM work. Nevertheless, it has been claimed that, because of the small hydraulic diameter of the flow channel in ECM, even optically smooth surfaces can act as rough surfaces [4].

2.8 Separation

In many ECM applications, flow over curved surfaces is more likely than flow over flat ones. With the former kind of flow, a phenomenon called ‘separation’ may arise.

Consider a fluid passing over a gently curved body, as indicated in Fig. 2.8. Upstream, between points A and B, the mainstream velocity just outside the boundary layer increases continually. From Bernoulli’s equation, it is clear that this rate of increase of velocity must be accompanied by a decrease in pressure. Downstream, beyond point C, a corresponding decrease in mainstream velocity and increase in pressure take place. That is, at a stage between B and C, the fluid within the boundary layer is affected by a pressure which increases in the direction of flow, normally called the ‘adverse pressure gradient’. Consequently, the fluid velocity may be reduced. If the velocity eventually becomes zero and then reverses direction, the boundary layer will be ‘separated’ from the body.

A necessary condition for separation can be derived from Prandtl’s boundary layer equations; in terms of Fig. 2.8, x is now regarded

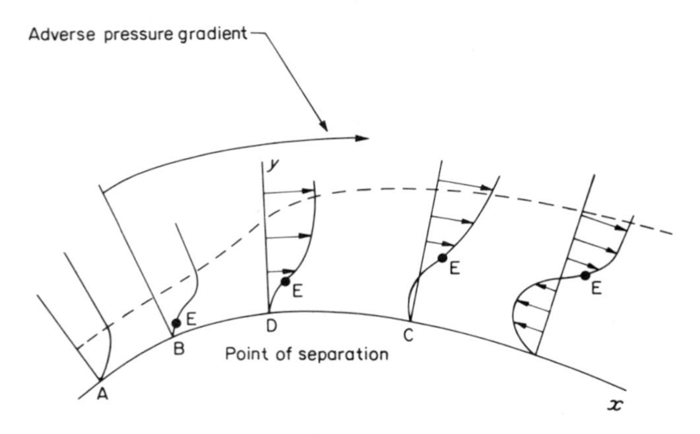

Fig. 2.8 Development of separation

as a curvilinear coordinate along the line of the boundary, and y is the normal distance from the boundary.

Suppose that separation starts at point D, at which the velocity and its variation in the y direction are zero. That is, at D,

$$\left(\frac{\partial u}{\partial y}\right)_{y=0} = 0 \tag{2.55}$$

But, at D also, $u = v = 0$. So

$$\frac{1}{\rho_e}\frac{\mathrm{d}p}{\mathrm{d}x} = \nu\left(\frac{\partial^2 u}{\partial y^2}\right)_{y=0} \tag{2.56}$$

First, for $\mathrm{d}p/\mathrm{d}x < 0$ (flow conditions between A and B), we see that $\partial^2 u/\partial y^2 < 0$ near the boundary. In the direction towards the mainstream, the velocity approaches the local mainstream asymptotically, and accordingly, the rate of decrease of $\partial u/\partial y$ is also reduced. Then, near the edge of the boundary layer, $\partial^2 u/\partial y^2 < 0$. Therefore, the curvature of the velocity profile is always negative for decreasing pressure.

Next, for $\mathrm{d}p/\mathrm{d}x > 0$, $\partial^2 u/\partial y^2 > 0$. But, flow conditions near the edge of the boundary layer have not altered, and there $\partial^2 u/\partial y^2 < 0$. It is deduced that an inflexion point in the profiles must exist; its position is indicated by E in Fig. 2.8.

On the boundary at the point of separation, $\partial u/\partial y = 0$. Since the velocity increases continuously from zero at the boundary and eventually to some positive value adjacent to the boundary, we infer that $\partial^2 u/\partial y^2 > 0$. But, at the edge of the boundary layer, $\partial^2 u/\partial y^2 < 0$, so that a point of separation must be associated with an inflexion point, and hence with an adverse pressure gradient.

Finally, we add that separation can also occur if the entry to the flow channel is abrupt. Also, in the wake downstream from the separation point, eddies and vortices are generated. Some effects which they appear to have on surface finish in ECM are discussed in Chapter 4.

2.9 Utilisation of fluid dynamic principles in ECM

Considerations of flow phenomena in ECM are usually based on the principles outlined in the above sections. The purpose of this section is to show examples of their use.

An important feature required for accurate study of fluid flow in ECM is that of an inlet length to provide fully developed velocity profiles between the electrodes. From Section (2.3.2), this inlet length is about 30 (laminar flow) to 100 (turbulent flow) pipe diameters. Its inclusion means that the inlet pressure has to be increased, often quite drastically. In practice, this increase is sometimes not possible, and hydrodynamic inlet lengths are often omitted, particularly in industrial ECM operations.

But many experimental ECM apparatuses do carry inlet lengths. One system has been described [5] which provided linear flow-rates between 1 and 25 m/s between rectangular electrodes of side 0·53 by 3·17 mm (smaller dimension in the direction of flow). In the cell, the electrolyte was forced down a rectangular channel of 0·5 x 8 mm cross section and 120 mm length, and the flow channel carried a hydrodynamic inlet length of 77 hydraulic diameters to ensure a fully developed velocity profile at the electrodes. The maximum pressure at the cell inlet was about 10^3 kN/m^2 which yielded a volume flow-rate of 10^{-4} m^3/s through the cell.

The electrode dimensions in that work were made much smaller than those normally met in ECM in order that carefully controlled experiments could be performed. In another investigation [6] rectangular electrodes of size 102 by 12·55 mm have been used (larger dimension in the flow direction) with an inlet length of 78 hydraulic diameters incorporated into the cell to provide fully established flow. Although these electrode dimensions are more likely in ECM, the electrode gap width in that work was 3·175 mm, which is greater than that usually found. Nonetheless, some characteristic data can be obtained from it. First, the hydraulic mean diameter [Equation (2.4)] is calculated:

$$d_h = \frac{4A}{C} = \frac{4 \times (3{\cdot}175 \times 12{\cdot}55)}{2(3{\cdot}175 + 12{\cdot}55)} = 5{\cdot}06 \text{ mm}$$

Note that here the approximation (2.5) cannot be used. The Reynolds number, Re, can now be found from Equation (2.3), in which the characteristic length is given by the hydraulic diameter:

$$\text{Re} = \rho_e \frac{\bar{u} d_h}{\mu} = 7740$$

where $\rho_e = 1{\cdot}09$ g/cm^3 and $\mu = 1{\cdot}19$ cP. [In this calculation, a value for the average velocity has been found from the volumetric flow-rate Q, as suggested in Section 2.3.2: $\bar{u} = Q/A = (0{\cdot}0666 \times 10^{-3})/(39{\cdot}8 \times 10^{-6} = 1{\cdot}67$ m/s).] Since Re > 2300 [Equation (2.31)], the flow can be regarded as turbulent.

These values have been used to estimate the viscous pressure drop down the channel for turbulent flow [Equations (2.32), (2.33), and (2.34)]. The friction factor f is first found:

$$f = \frac{0{\cdot}3164}{\text{Re}^{1/4}} = 0{\cdot}0337$$

The pressure drop is now given by

$$\frac{fL\rho_e}{2d_h} \bar{u}^2 = 1{\cdot}03 \text{ kN/m}^2$$

where L is the electrode length (= 102 mm).

Two other calculations are of interest. First, the thickness of the laminar sub-layer is found from Equation (2.44b):

$$\delta_1 = 124\, R(\text{Re})^{-7/8} = 0{\cdot}12 \text{ mm}$$

where $R = d_h/2$.

The corresponding thickness of the turbulent boundary layer is [Equation (2.51)]:

$$\delta_t = 0{\cdot}376\, x^{4/5} \left(\frac{\nu}{U}\right)^{1/5}$$

$$= 3{\cdot}5 \text{ mm}$$

where δ_t is calculated at $x = L$ (= 102 mm), $\nu = 1{\cdot}091$ mm^2/s, $U(\equiv \bar{u}$ now) = 1·67 m/s.

For interest, a hypothetical laminar boundary layer thickness is calculated from the same data:

$$\delta_0 = 5\left(\frac{\nu x}{U}\right)^{1/2} = 1{\cdot}29 \text{ mm}$$

With this hypothetical case, the thickness of the laminar boundary layer is seen to be less than that of the turbulent layer.

Estimates have been made [7] of the laminar boundary layer thickness over the inlet region of a 'quasi-rectangular' channel with gap widths of 0·1 to 0·5 mm and channel length of 2 mm. Equation (2.25) has been used to obtain this thickness, a numerical coefficient of three instead of five being used. For velocities in the range 0 to 9 m/s and for $\nu \simeq 1$ mm^2/s the estimated thickness is always less than half the gap width even at the exit. It is deduced that fully developed flow does not occur anywhere along this length of channel.

In a practical ECM cell the gap width is usually much less than breadth and length of the electrodes, which latter quantities can be large (e.g. 100 mm). The flow is usually turbulent. It is often useful to have an idea of the minimum pressure required to maintain a required flow-rate down the channel. We can find this amount for flow down a rectangular channel from Equations (2.30) and (2.17b) (laminar flow) or (2.34) (turbulent flow).

Example

These formulae have been applied to a case in which electrolyte of density $\rho_e = 1{\cdot}088$ g/cm^3 and viscosity $\mu = 0{\cdot}876$ cP has been pumped at a volume flow-rate of $0{\cdot}98 \times 10^{-3}$ m^3/s between rectangular parallel electrodes of area $76{\cdot}2 \times 38{\cdot}1$ mm^2 their gap distance at inlet being 0·92 mm.

The mean velocity is calculated to be 27·96 m/s and the Reynolds number 64 000 with $d_h = 2h = 1{\cdot}84$ mm. The friction factor ($f = 0{\cdot}3164/\text{Re}^{1/4}$) is calculated to be 0·0199. The dynamic pressure, all of which is assumed to be lost at exit from the channel, is now estimated to be

$$0{\cdot}5\rho_e\bar{u}^2 = 420 \text{ kN/m}^2$$

The pressure needed to overcome friction effects is

$$\frac{1}{2}\frac{f\rho_e L\bar{u}^2}{d_h} = 350 \text{ kN/m}^2$$

Thus the total minimum pressure required to force the electrolyte down this channel is estimated to be 770 kN/m^2. In practice, the actual pressure drop without ECM was found to be 613 kN/m^2. The discrepancy can be attributed to a possible difference between the calculated friction factor and the experimental one, and to the neglect of other possible hydrodynamic factors (e.g. cavitation – see below – which might have affected the friction loss calculations). But the calculations do demonstrate their usefulness in providing a minimum value for the required pressure. With ECM the required pressure will be increased even more.

2.10 Multi-phase flow

The generation of gas at the electrodes, particularly cathodic hydrogen, and the dissolution of metal from the anode, mean that the fluid in the machining gap in ECM is of a multi-phase, rather than a single-phase, nature.

In most investigations of fluid flow in ECM, only the gas bubbles are assumed to have appreciable effect on the nature of the flow, which therefore can be regarded as two-phase. The pattern of two-phase flow is determined by the bubble (or void) distribution, an understanding of which usually also demands knowledge of the velocity and shear distributions of the flow. Such information is often difficult to find. Nonetheless, with two-phase flow, certain typical phenomena can arise, and in this section some which are relevant to ECM are briefly described.

2.10.1 Effect on pressure drop

Although this aspect of two-phase flow on the ECM process is also considered in Chapter 5, one feature can usefully be introduced here, namely, the effect on the pressure drop for rectangular channel flow. In Fig. 2.9 are shown pressure drop measurements for a 10% NaCl solution flowing down a channel of length 152 mm, breadth 25 mm, and (gap) width 3·2 mm. The pressure drop without ECM can be estimated from Bernoulli's equation and the Blasius relation given above (for turbulent flow). For flow with ECM, a higher pressure drop is required to maintain the same volume flow-rate of solution, and this increase in pressure drop itself becomes greater at the higher rates of ECM, and becomes less with the higher pressures at outlet to the gap.

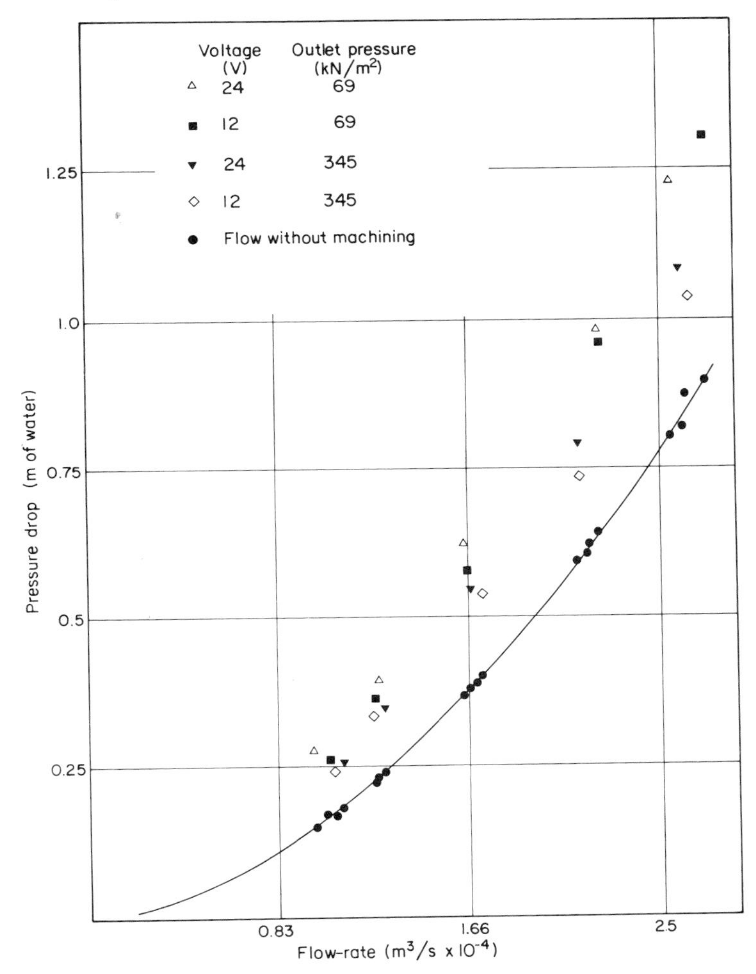

Fig. 2.9 Pressure drop without ECM and with ECM (after Clark [6])

2.10.2 *Choking flow*

The presence of gas in the electrolyte solution in ECM means that 'choking', a phenomenon associated with compressible fluid flow, could arise. The characteristics of choking can be illustrated from considerations of one-dimensional flow down a duct.

First, consider the flow of an incompressible fluid which starts from rest in a large reservoir of high pressure. Since the density of the fluid is constant and if the mass flow-rate is also assumed to be constant, the velocity will rise continuously as the channel area

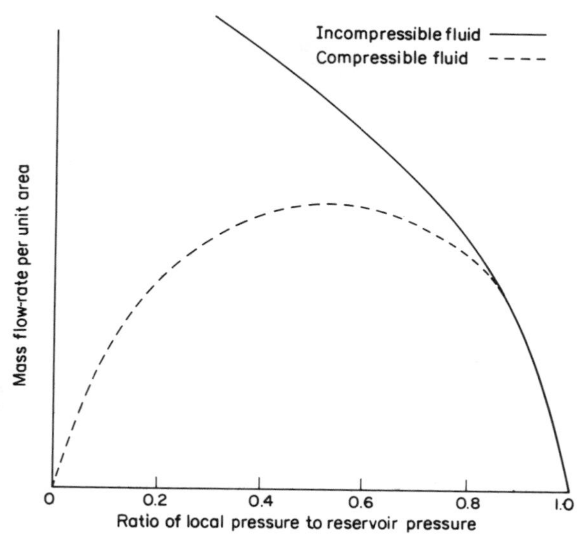

Fig. 2.10 Variation of mass flow rate per unit area with pressure for an incompressible and a compressible-fluid (e.g. air), expanded isentropically

decreases and as the pressure falls to zero. The mass flow-rate per unit area will also rise continuously with falling pressure. The maximum velocity, and hence mass flow-rate, can, of course, be found from Bernoulli's equation.

On the other hand, for a compressible fluid, it can be shown that as the pressure falls the mass flow-rate per unit area first increases. But the mass flow-rate, after reaching a maximum value, decreases as the density decreases. This maximum is determined only by conditions in the reservoir and by the smallest area of the duct. This condition is referred to as *choking*. This critical mass flow-rate cannot be increased further by further change in differential pressure along the length of the channel. The variation of mass flow-rate per unit area with pressure for both an incompressible and a compressible fluid is shown in Fig. 2.10.

2.10.3 Cavitation

As the electrolyte solution flows between the electrodes during ECM, its temperature will rise owing to, for example, the passage of current and the pressure drop along the flow channel. This increase can lead to the growth of gas-filled bubbles (cavities). If the growth of the bubbles is caused by temperature increase, boiling occurs. If their growth is due to pressure reduction (at constant temperature) 'cavitation' is said to have occurred.

Studies of cavitation relate to (i) the formation of bubbles in the liquid caused by reductions in pressure below certain critical values, determined by the physical condition of the liquid, and (ii) the collapse of bubbles due to an associated increase in pressure. Cavitation behaviour is known to take place either in the bulk of the liquid or at its interface with a boundary, and the presence of cavities can lead to displacement of the liquid phase of the solution, causing alteration in the flow pattern. A further effect can be damage to the channel walls. Suppose that, during ECM, the pressure conditions and the configuration of the electrodes lead to the formation of cavities which are later swept by the flow into regions of higher pressure where the cavities collapse. If the point of collapse of cavity is in contact with either of the electrodes, the wall of the electrode will receive a blow. The consequence of successive blows may be damage to the electrode material.

In ECM cavitation is more likely to arise at a location where an abrupt change occurs in the direction of flow, associated with which is a sudden decrease in the local pressure. The problem has been the subject of a number of investigations [8, 9]. In one study of electrochemical die-sinking, the electrolyte was pumped through a central nozzle and then outwards in a radial direction. Calculations, based on Bernoulli's equation and the equation for the pressure drop caused by viscous effects, showed that pressures near the nozzle inlet could be less than the vapour pressure of the electrolyte, indicating that cavitation was likely in that region. This was confirmed by experiment. It was also found that the onset of cavitation could be prevented by the application of a sufficiently high outlet or 'back' pressure. Cavitation was also eliminated when the electrolyte was caused to flow in the reverse direction. The shape of the nozzle also played an important role: a sharp-cornered nozzle required a higher limiting outlet pressure to eliminate cavitation than did a smoothly tapered nozzle. (No quantitative information is available to support this observation. Presumably a sharp-cornered nozzle caused the flow direction to change abruptly, with a correspondingly abrupt decrease in pressure. These conditions would then be more favourable to cavitation.)

Figure 2.11 shows another configuration of electrodes used to study cavitation in ECM [10]. The electrodes are plane and parallel

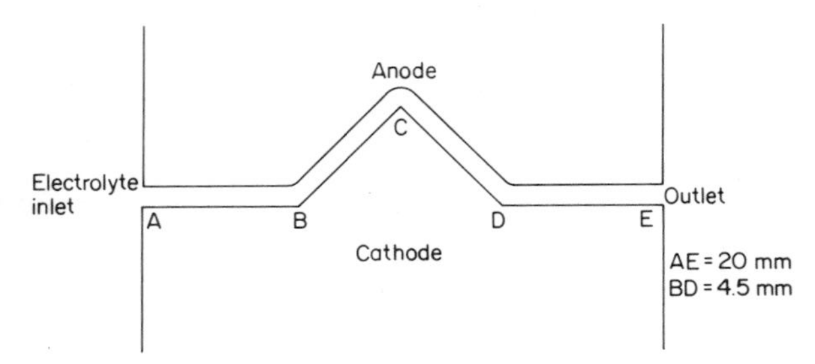

Fig. 2.11 Configuration of electrodes for investigation of cavitation in ECM (after Ito and Seimiya [10])

along the length AB whilst along BCD the cathode is in the shape of a right-angled wedge. A 10% (w/w) NaCl solution was forced down the gap at Reynolds numbers ranging from 8 000 to 20 000, the pressure at inlet being 1570 kN/m^2 and the outlet pressure being varied from 980 to 1170 kN/m^2.

Cavitation bubbles are seen as the white layer at the top of the wedge portion of the cathode in Fig. 2.12. The amount of cavitation

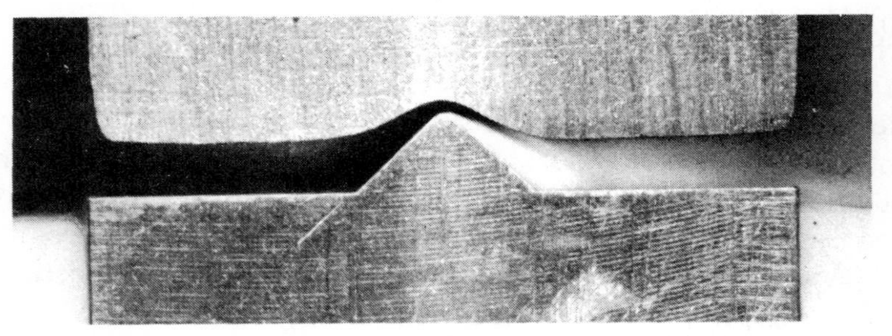

Fig. 2.12 Cavitation during ECM (voltage = 15 V; feed-rate = 0·02 mm/s; outlet pressure = 294 kN/m^2) (By permission of S. Ito and K. Seimiya [10])

could again be reduced by increasing the outlet pressure. For the conditions given above, cavitation was eliminated when the outlet pressure was about 780 kN/m^2.

Cavitation will be discussed later in connection with surface finish and limitations on the rate of ECM.

References

1. See, for example, Schlichting, H., *Boundary Layer Theory*, McGraw-Hill, New York (1968).
2. See, for example, Chin, D. T., *J. Electrochem. Soc.* (1972) **119**, No. 8, 1043.

3. Knudsen, J. G. and Katz, D. L., *Fluid Dynamics and Heat Transfer*, McGraw-Hill, New York (1958).
4. Tobias, C. W., Paper presented at First Int. Conf. on ECM, Leicester University, March 1973.
5. Landolt, D., Muller, R. H. and Tobias, C. W., *J. Electrochem. Soc.* (1969) **116**, No. 10, 1384.
6. Clark, W. G., Ph.D. Thesis, Strathclyde University, Glasgow (To be submitted).
7. Kruissink, C. A., Paper presented at First Int. Conf. on ECM, Leicester University, March 1973.
8. Chikamori, K., Ito, S. and Sakurai, F., *Bull. Jap. Soc. Prec. Eng.* (1968) **2**, No. 4, 318.
9. Ito, S., Chikamori, K. and Sakurai, F., *J. Mech. Lab. Japan* (1966) **12**, No. 2, 37.
10. Ito, S. and Seimiya, K., *Proc. 12th Int. Conf. Mach. Tool Res.*, Macmillan (1972) p. 259.

Bibliography

Knapp, R. T., Daily, J. W. and Hammitt, F. G., *Cavitation*, McGraw-Hill, New York (1970).

Knudsen, J. G. and Katz, D. L., *Fluid Dynamics and Heat Transfer*, McGraw-Hill, New York (1958).

Schlichting, H., *Boundary Layer Theory*, McGraw-Hill, New York (1968).

Shames, I. H., *Mechanics of Fluids*, McGraw-Hill, New York (1962).

Tong, L. S., *Boiling Heat Transfer and Two-Phase Flow*, Wiley, New York (1965).

Wallis, G. B., *One Dimensional Two-Phase Flow*, McGraw-Hill, New York (1969).

CHAPTER THREE

Basic Electrochemistry

Although those aspects of fluid dynamics given in the previous chapter are useful in outlining the patterns of flow likely to occur in ECM, clearly an understanding of the process also demands some knowledge of basic electrochemistry.

Relevant information can be derived from studies of the structure and properties of electrolytes, since these media largely determine the rates and types of reactions in ECM. Upon that framework can be built an appreciation of the interactions which arise when a single metal is placed in an electrolyte, and of those other equilibrium phenomena associated with an electrochemical cell consisting of two metals dipped in an electrolyte and externally connected.

Studies of these so-called reversible processes certainly provide a guide to the likelihood of certain reactions occurring in practice. But in ECM the only anode reaction which is wanted is metal dissolution, and that is achieved by an externally applied potential difference between the two electrodes. The process is now said to be irreversible, and the electrodes *polarised.* With an irreversible process are closely associated overpotentials, which are a measure of its departure from reversibility. Investigations of overpotentials in ECM require some of those aspects of fluid dynamics which were given in Chapter 2.

3.1 Basic assumptions and definitions

(*a*) *Electrolyte conductivity*

In the first chapter, the idea of the specific conductivity of the electrolyte was introduced. The magnitude of the conductivity is determined by the types and numbers of the ions present in the electrolyte. The different ions may also carry different quantities of electric charge. It is useful, therefore, to compare the conductivities of different electrolytes by reference to concentration, a quantity which can be applied to all electrolytes. Concentrations are often calculated from either the weight of the solute per unit weight (w/w), or weight per unit volume (w/v), of solution, or the volume of the solute per unit volume (v/v) of solution.

Of these measures of concentrations, a commonly used quantity is that of gram-molecules, or moles, of solute per litre of solution. (A gram-molecule is the molecular weight in grams.) Thus a one-molar (1M) solution is one which contains one gram-molecule of solute per litre of solution; a 2M solution contains two gram-molecules, etc. For example, a 1M NaCl solution has 58·5 g of NaCl per litre, since the atomic weights of Na and Cl are 23 and 35·5 respectively. The gram-equivalent of solute per litre of solution is another useful measure of concentration. (A gram-equivalent is the equivalent weight in grams.) A solution with one gram-equivalent of solute per litre of solution is a one-normal (1N) solution, For example, a 1N H_2SO_4 solution has 49 g of H_2SO_4 per litre of solution, since the respective atomic weights are 1, 32, and 16.

Concentrations of electrolytes so expressed provide a common base for the comparison of the conductivities of different electrolytes. In particular, the molar conductivity, Λ_m, and equivalent conductivity, Λ_c, are useful:

$$\Lambda_m = \frac{\text{specific conductivity at concentration } C \text{ of solution}}{\text{concentration of solution in moles per ml of solution}}$$

$$\Lambda_c = \frac{\text{specific conductivity at concentration } C \text{ of solution}}{\text{concentration of solution in g-equivalents per ml of solution}}$$

To establish the relationship between the molar and equivalent conductivities, we introduce the electrochemical valency n_e.

Suppose that an electrolyte of general formula $P_x Q_y$ is electrically neutral and that, on ionisation, each molecule forms xP^{z+} cations and yQ^{z-} anions. Then each g-molecule of $P_x Q_y$ contains $xz_+ (= yz_-)$ g-equivalents. (The condition of electrical neutrality gives that $xz_+ = yz_- = n_e$; for instance, for NaCl, $x = 1$, $z_+ = 1$, $y = 1$, and $z_- = 1$, so $n_e = 1$; whilst with Na_2SO_4, $x = 2$, $z_+ = 1$, $y = 1$, $z_- = 2$, so $n_e = 2$.)

Thus, the molar conductivity Λ_m at concentration C g-molecules/l is

$$\Lambda_m = \frac{1000\ \kappa_e}{C} \tag{3.1}$$

whilst the corresponding equivalent conductivity Λ_c is

$$\Lambda_c = \frac{1000\ \kappa_e}{n_e C} \tag{3.2}$$

(b) Degree of dissociation

It has been assumed so far that if a substance dissolves in water, it dissociates completely into ions. This is not necessarily the case. It becomes useful, therefore, to define the 'degree of dissociation', α, of an electrolyte, that is, the fraction of the solute which is dissociated into ions that are free to carry current at a given concentration. On the basis of this definition, a rough division of electrolytes can be made: 'strong' electrolytes, which become greatly dissociated for concentrations ranging from very low to high values (α always about unity), and 'weak' electrolytes, for which α tends to unity at very low, limiting concentrations, and reduces to about zero at high concentrations. (It is important to distinguish between 'dissociation' and 'ionisation'. The former term applies to ions which are free to carry current, whilst the latter refers to the total number of ions present and not to their ability to carry current. Electrostatic attraction between ions of opposite sign can lead to the temporary formation of ion-pairs which behave as non-ionised molecules and so are unable to carry current. Consequently, a wholly ionised electrolyte may not become completely dissociated.) Weak electrolytes, of which weak organic acids are typical examples, are seldom used in ECM except as additives to the main solution to improve the quality of machining. Strong electrolytes are used as the main electrolytes in ECM; a common example is NaCl solution.

The degree of dissociation of an electrolyte can be related to Kohlrausch's observation that, at very low concentrations of an electrolyte, each ion makes a definite contribution towards the equivalent conductivity of the electrolyte, irrespective of the nature of the other ion with which it is associated in the solution. Before this relationship is established, the notion of ionic mobility must be introduced.

(*c*) *Ionic mobility*

We have already seen that it is the motion of the ions which makes possible the passage of current through the electrolyte. The rate of ionic movement is termed the ionic 'mobility'. Since the current is equal to the quantity of electrical charge passed per second, and since the cations and anions carry this charge, the sum of the velocities of the anions and cations is directly proportional to the current. From Ohm's law, the potential difference between the electrodes is also directly proportional to the sum of the ionic velocities. When the potential gradient is 1 V/cm, the ionic velocity is termed the 'absolute ionic mobility', the symbols u_+ and u_- being used for the cations and anions respectively. Some typical absolute ionic mobilities at 25°C are given in Table 3.1.

Table 3.1 Absolute ionic mobilities

Cation	Hydrogen	Potassium	Sodium	
Mobility (m/s) x 10^{-6}	36·2	7·6	5·2	
Anion	Hydroxyl	Sulphate	Chloride	Nitrate
Mobility (m/s) x 10^{-6}	20·5	8·3	7·9	7·4

(*d*) *Transport numbers*

Although the sum $u_+ + u_-$ is proportional to the current, the amount carried by each ion is proportional to its own velocity. The fraction of the total current carried by each ion type is called the 'transport number':

$$t_+ = \frac{u_+}{u_+ + u_-} \quad (3.3a)$$

and

$$t_- = \frac{u_-}{u_+ + u_-} \quad (3.3b)$$

Here, t_+ and t_- are the transport numbers of the cations and anions, respectively, in the electrolyte. Transport number values are affected by temperature and ionic concentration and usually have a limiting value at very low concentrations.

Measurements of transport numbers often reveal that ions in solution have become attached to one or more water molecules, i.e. they have become hydrated. Those water molecules in the 'primary hydration sheath' close to the ion are usually tightly bound to the ion. The others in the 'secondary hydration sheath' are more easily removed.

(*e*) *Relationship between electrolyte conductivity and concentration*

Now consider an electrolyte P_xQ_y of concentration C g-mole/1, of electrochemical valency n_e, and of degree of dissociation α. For a potential drop of 1 V across a cube of side 1 cm of the electrolyte, Ohm's law gives $I = \kappa_e$ where I is the current flowing across the cube. The quantity of anions, or cations, present in the cube is $\alpha n_e C/1000$ g-equivalents. Each cation and anion g-equivalent contains N/z_+ and N/z_- ions respectively, where N is Avogadro's number; as usual, z_+ and z_- are the cationic and anionic valency. (Avogadro's number is the number of atoms in one gram-atom, 6×10^{23}.) Therefore, the number of cations and anions in the cube are $\alpha n_e CN/1000z_+$ and $\alpha n_e CN/1000z_-$ respectively. Now suppose that current I is passed for 1 s. This current causes the discharge at the electrodes of a number $\alpha n_e CNu_+/1000z_+$ of cations (occupying a volume u_+), and of a number $\alpha n_e CNu_-/1000z_-$ of anions (occupying a volume u_-). The corresponding electric charges carried by the groups of ions are $e\alpha n_e CNu_+/1000$ and $e\alpha n_e CNu_-/1000$, where e is the electronic charge. Since the total rate of charge is I, we have

$$I = \frac{e\alpha n_e CN}{1000}(u_+ + u_-) = \kappa_e$$

From Equation (3.2)

$$\kappa_e = \frac{n_e C\Lambda_c}{1000}$$

and, since $eN = F$, Faraday's constant,

$$\Lambda_c = \alpha F(u_+ + u_-) \tag{3.4}$$

It is customary to denote the quantities Fu_+ and Fu_- by λ_+ and λ_- respectively, so that

$$\Lambda_c = \alpha(\lambda_+ + \lambda_-)$$

At very low concentration, $\alpha = 1$, then

$$\Lambda_0 = (\lambda_+)_0 + (\lambda_-)_0$$

where here the suffix 0 denotes values at very low concentration.

These equations demonstrate that cations and anions contribute independently to the equivalent conductivity.

(*f*) *Onsager's equation*

Because of the electrical attraction between positive and negative ions, an ion may be regarded as being surrounded by a centrally symmetric ionic atmosphere whose total charge is equal and opposite to that of the central ion. When an electric field is applied, the ions are made to move. The central ion moves in one direction and during the finite time required by the atmosphere to readjust to the new conditions, the central ion is unsymmetrically positioned within the atmosphere, and the electrostatic attraction between it and its atmosphere reduces the motion of the central ion. In addition, the applied field makes the ionic atmosphere (carrying its customary molecules of water of hydration) move in the opposite direction to that of the central ion; as a result, the ion movement is retarded further by the dragging effect of the atmosphere.

On the basis of investigations of these phenomena by Debye and Hückel, Onsager put forward the following equation for a strong electrolyte which produces two univalent ions in solution (e.g. KCl or NaCl):

$$\Lambda = \Lambda_0 - (A + B\Lambda_0)\sqrt{C} \qquad (3.5)$$

where Λ_0 is the limiting conductivity, A and B are constants for the given solvent and depend only on temperature, and C is the concentration of the solution in g-mole/l or g-equiv/l. The 'A' term describes the electrophoretic effect due to the dragging of the atmosphere, whilst the expression given by 'B' is caused by the asymmetrical positions of the central ion and atmosphere. The equation demonstrates that the equivalent conductivity decreases from the limiting value with increasing concentration.

For water at 25°C, $A = 60{\cdot}2$ and $B = 0{\cdot}229$, so that for a strong aqueous electrolyte of type PQ, with both P^+ and Q^- ions univalent, we have

$$\Lambda = \Lambda_0 - (60{\cdot}2 + 0{\cdot}229\Lambda_0)\sqrt{C}$$

that is, Λ decreases with the square root of C.

Onsager's equation is accurate for electrolyte concentrations up to about 2×10^{-3} g-equiv/l. The corresponding behaviour of Λ at higher concentrations has yet to be fully established, mainly owing to difficulties in the analysis of the ionic charge distribution in very concentrated solutions. Highly concentrated electrolytes are, of course, widely used in ECM. That the dependence of conductivity on concentration for such solutions is so complex is representative of the many problems encountered with the ECM process.

(*g*) *Activity*

Conditions of electrochemical equilibrium are really dependent not only on the concentrations of ions present in solution, but also upon the extent to which the ions react with each other and with the water molecules in the solution. Because of this overall dependence concentration is not really a true indication of the ability of ions to determine the properties of an electrolyte. A more accurate description can be obtained on the replacement of the 'concentration' term by an 'activity', a, which takes into account the interactions between an ion and its surroundings. The relationship between the two can be conveniently expressed as

$$a = fC \tag{3.6}$$

where f is the activity coefficient. At very low concentrations, where there is negligible ionic interaction, the activity coefficient approaches unity. Since the amount of ionic interaction varies with concentration, some variation of activity coefficient with concentration can be expected. Table 3.2 shows the variation of f with concentration for NaCl electrolyte.

(*h*) *pH values*

Electrolytes can also be classified into acidic, neutral, or alkaline types. This classification is usually defined by the pH of the electrolyte, that is the negative logarithm of the hydrogen ion concentration, C_H, expressed in g-ion per litre:

$$\text{pH} = -\log_{10} C_H \tag{3.7}$$

Note: the more formal definition of pH requires that the concentration term be replaced by activity. But measurement of activity is rendered difficult by experimental problems, and in practice concentration is the preferred term.

Table 3.2 Variation of mean activity coefficient with concentration of NaCl at 25°C

Concentration (g mole/litre water)	Mean activity, f
0·01	0·902
0·1	0·778
0·5	0·681
1·0	0·657
1·5	0·656
2·0	0·668
3·0	0·714
5·0	0·874

Consider next the dissociation of water:

$$H_2O \rightarrow H^+ + OH^-$$

The actual amount of dissociation of water is very small; in fact, pure water contains only 10^{-7} g-ion of hydrogen per litre at 25°C, and the concentration of hydrogen ions is equal to that of hydroxyl ions. The water is said to be *neutral*; from the definition (3.7), its pH value is 7. This value of pH represents the neutral point in a scale whereby solutions can be classified by their degree of acidity. If the hydrogen ion concentration of a solution is greater than 10^{-7}, the pH is less than 7 and the solution is *acidic.* If the solution has a pH of 7, it is *neutral.* When the hydrogen ion concentration is less than 7, the pH value exceeds 7, and the solution is *alkaline.* Later, the electrode reactions which occur in these types of electrolytes will be discussed. But at this stage it can be noted that neutral electrolytes (e.g. NaCl, $NaNO_3$) are most commonly used in ECM, although for some applications, e.g. drilling, acid electrolytes are preferred.

(*i*) *Solubility product*

When a saturated solution of an electrolyte $P_x Q_y$ is in contact with the solid solute, an equilibrium exists between the ions in the solution and the solid. For a simple electrolyte PQ ($x = 1, y = 1$), the equilibrium may be represented by

$$PQ(s) \rightleftharpoons P^+ + Q^-$$

The equilibrium constant, K, is given by

$$K = \frac{[a_{P^+}] \times [a_{Q^-}]}{[a_{PQ}]}$$

where the square brackets represent activities (or, alternatively, concentrations). If the activities of a pure liquid and a pure solid

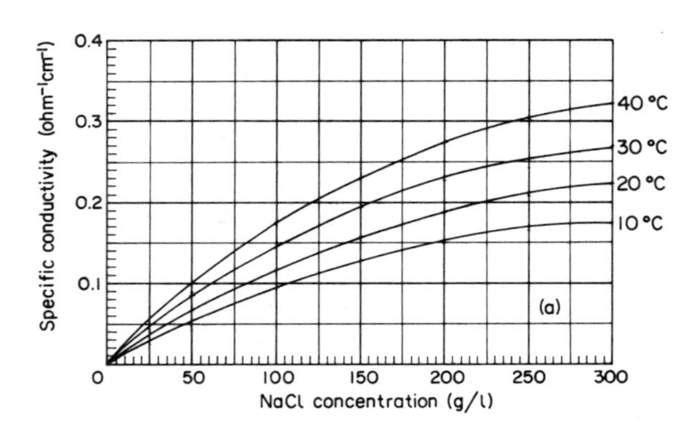

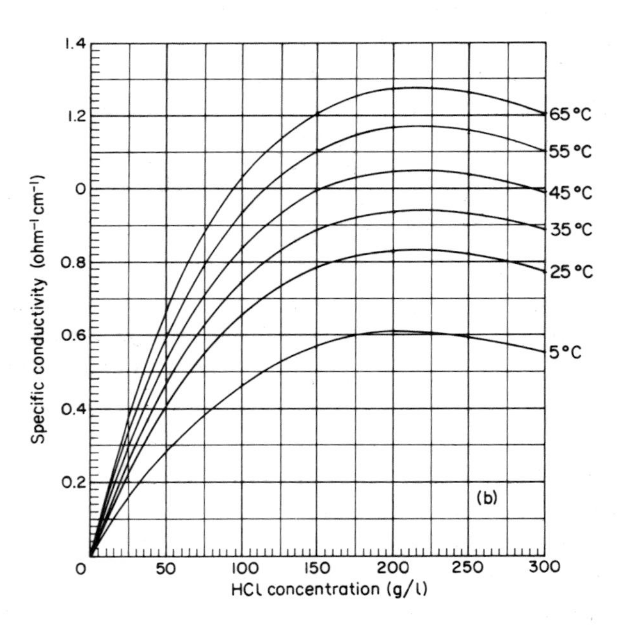

Fig. 3.1 Specific conductivity as a function of concentration and temperature for (a) NaCl and (b) HCl electrolytes

are taken to be unity, the above expression can be written

$$K_s = [a_{P^+}] \times [a_{Q^-}]$$

K_s is known as the activity solubility product.

A significant observation from this result is that the product of the activities of the ions P^+ and Q^- in a saturated solution of PQ must be constant for a given temperature. This condition holds irrespective of whether ions arise solely from the substance PQ or whether any contribution is made to their activities by the presence of other electrolytes in the solution.

3.2 Properties of ECM electrolytes

Based on the information given in the preceding section, several relevant general properties of electrolytes can be deduced. First, since the ionic mobility increases with temperature, the electrical conductivity of the electrolyte should have a similar temperature dependence (because of the relationship between conductivity and mobility, e.g. Equation (3.4).) Secondly, we expect some variation in conductivity with concentration, bearing in mind that concentrations of ECM electrolytes greatly exceed those for which the Debye–Hückel–Onsager theory is valid.

In Fig. 3.1(a) the variation of specific conductivity with concentration and temperature for NaCl solution is shown. Corresponding results for HCl solution are presented in Fig. 3.1(b). With the former electrolyte, the conductivity increases with temperature and with concentration up to the limit of solubility of the salt. With the HCl solution, the conductivity increases with concentration to a maximum value and then decreases. Data for a range of electrolytes and for conditions encountered in ECM can be found elsewhere [1, 2].

3.3 Equilibrium electrode potentials

3.3.1 Electrical double layer

When a metal is placed in an electrolyte solution, an equilibrium potential difference usually becomes established between the metal and solution. This potential difference arises from the transfer into solution of metal ions and the simultaneous discharge of ions from the solution. Equilibrium is reached when the electrons left in the metal contribute to the formation of a layer of ions whose charge is equal and opposite to that of the cations in solution at the interface. The positive charges in the solution and negative charges in the metal form the *electrical double layer.* Although the structure

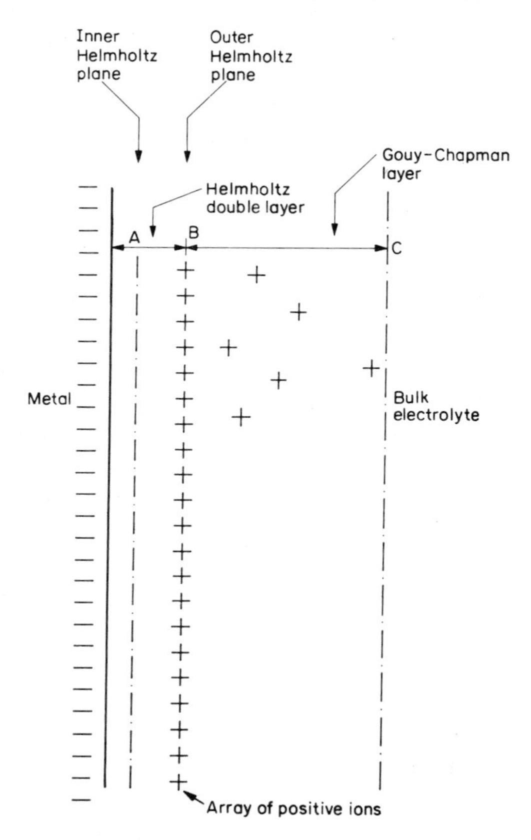

Fig. 3.2 The electrical double layer

of the double layer is complicated, the simple representation in Fig. 3.2 highlights several characteristic regions. Over the inner Helmholtz plane A, which lies very close to the metal, unsolvated charges which do not have molecules of water attached to them may be adsorbed. Beyond A is the outer Helmholtz plane, B, along which lies an array of the positive charges. Extending further into the solution is the Gouy–Chapman layer, C, which is more diffuse and mobile. Over that layer the potential drop is non-linear. The bulk electrolyte with its usual properties is situated outside the

Gouy–Chapman layer. The Helmholtz double layer has a minimum thickness of about 2×10^{-7} mm. The approximate thickness of the Gouy–Chapman layer is 10^{-3} mm. The transfer of ions clearly will cease when the energy required for an ion to dissolve is less than the work necessary to pass it across the double layer. The reverse case also holds for the formation of a double layer owing to the deposition of ions on the metal.

3.3.2 *Nernst equation*

It will be of use to derive an expression for the equilibrium potential difference between the metal and its ion in solution [3].

Suppose that the atoms of the metal can be dissociated into metal ions and free electrons:

$$M \rightleftharpoons M^{z+} + ze$$

At the metal surface, ions are held by lattice forces, and they require an activation energy W_1 to pass the metal–solution interface into solution. In the solution, ions are held by forces of hydration; they need an activation energy W_2 to cross the interface in the opposite direction. From Maxwell's energy distribution law for the reacting particles, it can be shown that the number of particles, n, which have sufficient energy to cross the interface can be related to the total number of particles, n_t, by an expression of the form $n = n_t \exp(-W/kT)$. The respective rates of crossing from metal to solution, and vice versa, can then be deduced:

$$R_1 = K_1 \exp(-W_1/kT)$$

$$R_2 = CK_2 \exp(-W_2/kT)$$

K_1 and K_2 are characteristic constants for the metal and solution which are independent of concentration, C is the free metal ion concentration in solution, T is the temperature, and k is Boltzmann's constant.

If R_1 is greater than R_2, ions will dissolve at a rate greater than that at which they deposit, and vice versa if R_1 is less than R_2. For either condition, an electrical double layer is established. A change in electrical potential difference takes place across the double layer, so that the energy required by an ion to pass from metal into solution is reduced to $(W_1 - zeE_1)$, and the energy required by an ion to cross from solution to metal is increased to $(W_2 + zeE_2)$. Here, E_1 is the potential drop associated with the reduction of the energy barrier at the metal side of the interface, and E_2 is the corresponding potential drop on the solution side.

The rates at which ions cross in both directions now are

$$R_1' = K_1 \exp[-(W_1 - zeE_1)/kT] \tag{3.8}$$

$$R_2' = CK_2 \exp[-(W_2 + zeE_2)/kT] \tag{3.9}$$

In equilibrium, $R_1' = R_2'$. Thus,

$$E = E_1 + E_2 = \frac{W_1 - W_2}{ze} + \frac{kT}{ze} \ln\left(\frac{CK_2}{K_1}\right)$$

Since $RT/zF = kT/ze$, and putting

$$E_0 = \frac{W_1 - W_2}{ze} + \frac{kT}{ze} \ln\left(\frac{K_2}{K_1}\right)$$

we obtain the Nernst equation:

$$E = E_0 + \frac{RT}{zF} \ln C \tag{3.10}$$

R is the Gas Constant and E_0 is the normal electrode potential. (More correctly, of course, the concentration term in the Nernst equation should be replaced by the activity a.) Alternative derivations of the Nernst equation are available elsewhere [4].

For a more general cell reaction at reversible equilibrium, we have

$$M_1 + M_2^{z+} \rightleftharpoons M_1^{z+} + M_2$$

The general expression for this reaction, corresponding to the Nernst equation, is written

$$E = E_0 + \frac{RT}{zF} \ln \frac{[a(M_1^{z+})]\,[a(M_2)]}{[a(M_1)]\,[a(M_2^{z+})]} \tag{3.11}$$

Example

When C is 1 g-ion/litre, $E = E_0$, and the departure of C from this value causes a change in potential:

$$(E - E_0) = \frac{RT}{zF} \ln C$$

$$= 2{\cdot}303 \frac{RT}{zF} \log_{10} C$$

When $T = 298$ K (= 25°C), $2{\cdot}303RT/F = 0{\cdot}059$ V. For $z = 1$, and an increase in C to 10 g-ion/litre, $(E - E_0) = 59$ mV, whilst for $z = 2$, the same change in C gives $(E - E_0) = 29{\cdot}5$ mV.

3.3.3 Free energy of reaction

The equilibrium rates of ion movement, R_1' and R_2', can also be represented by the *exchange current density*, J_0. This current density is usually of the order of 10^{-1} to 10^{-5} A/cm². It can be expressed as

$$J_0 = \Omega_1 \exp(-\Delta G_1/RT) - \Omega_2 \exp(-\Delta G_2/RT) \qquad (3.12)$$

where Ω_1 and Ω_2 are characteristic parameters for the dissolution and deposition reactions, R is the Gas Constant, and T is the temperature. ΔG_1 and ΔG_2 indicate the respective minimum free energies which atoms must possess to ionise to the solution and which ions must have to discharge from the solution. As shown in Fig. 3.3, the difference in the free energies ΔG_1 and ΔG_2 is ΔG, the free energy change for the reaction. ΔG is related directly to the reversible electrode potential, E, by

$$\Delta G = -zEF \qquad (3.13)$$

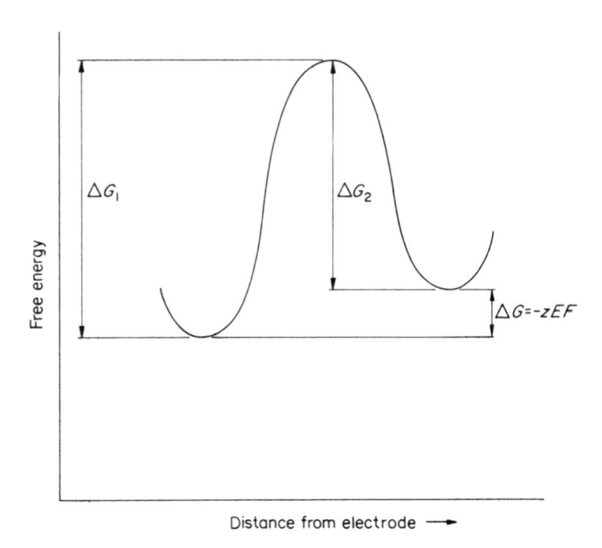

Fig. 3.3 Free energy–distance diagram for equilibrium conditions

3.3.4 Standard electrode potentials

Consider now two metals M_1 and M_2 placed in an electrolyte which contains their ions and which are externally linked at the junction M_1/M_2. Current is assumed to flow round the completed circuit in the direction of electron flow shown in Fig. 3.4. The potential drop in the circuit due to extraneous factors (e.g. ohmic drop in the bulk of the solution and along the electrodes' lengths) is assumed to be negligible. In the metal M_1 electrons move towards the M_1/electrolyte interface as the M_1 ions discharge from the metal into the solution. At the metal M_2, an equivalent quantity of M_2 ions discharge on it causing electrons to flow away from the M_2/electrolyte interface round the circuit. At the junction M_1/M_2, the electrons must transfer from M_2 to M_1. The energy required for this transfer is usually obtained from the *thermionic work function* ψ. Thus if the energy required to transfer the electron from the metal into free space is ψ_1 and if ψ_2 electron-volts are released by the transfer of an electron from free space into M_2, the transfer of an electron from M_1 to M_2 is identified with an energy change of $(\psi_1 - \psi_2)$ electron-volts. (Note: an electron-volt is the change in energy of an electron passing through a potential drop of 1 volt.) This number of electron-volts $(\psi_1 - \psi_2)$ is usually referred to as the potential difference at the junction of M_1 and M_2, or contact potential of M_1 and M_2.

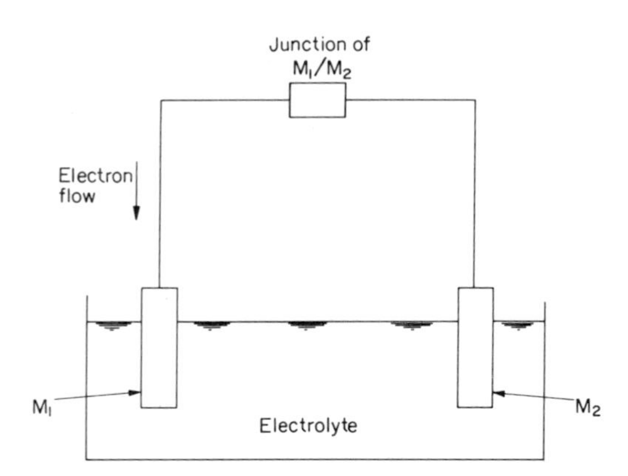

Fig. 3.4 Metals M_1 and M_2 in an electrolyte containing their ions

It is convenient to have a standard electrode whose reversible potential is made arbitrarily zero and against which the potentials of other electrodes can be measured. The *hydrogen electrode* is an accepted standard: it comprises a rod of platinum covered with platinum black saturated with hydrogen gas at atmospheric pressure. Electrode potentials based on this zero are said to refer to the *hydrogen scale.* However, in experimental work, it is often more

Table 3.3 Standard electrode potentials with reference to the hydrogen electrode at 25°C.

Electrode	Reaction	Volts
Lithium	$Li \rightarrow Li^+$	−2·96
Potassium	$K \rightarrow K^+$	−2·92
Sodium	$Na \rightarrow Na^+$	−2·72
Magnesium	$Mg \rightarrow Mg^{++}$	−2·38
Aluminium	$Al \rightarrow Al^{+++}$	−1·67
Zinc	$Zn \rightarrow Zn^{++}$	−0·76
Iron	$Fe \rightarrow Fe^{++}$	−0·44
Tin	$Sn \rightarrow Sn^{+++}$	−0·34
Nickel	$Ni \rightarrow Ni^{++}$	−0·25
Lead	$Pb \rightarrow Pb^{++}$	−0·12
Hydrogen	$H \rightarrow H^+$	0
Saturated calomel		+0·24
Copper	$Cu \rightarrow Cu^{++}$	+0·34
Silver	$Ag \rightarrow Ag^+$	+0·80
Oxygen	$2H_2O \rightarrow O_2 + 4H^+$	+1·23
Chlorine	$2Cl \rightarrow Cl^{++}$	+1·36
Gold	$Au \rightarrow Au^{+++}$	+1·50

convenient to use another standard electrode. The calomel electrode is a common example. It consists of a pool of mercury covered with calomel (mercurous chloride) and immersed in a solution of potassium chloride which acts as the electrolyte. A platinum wire allows electrical contact with the mercury. The electrode potential depends mainly on the concentration of the potassium chloride; for example, at saturation its potential on the hydrogen scale is about 0·24 V at 25°C. In Table 3.3 is given a range of standard electrode potentials, referred to the hydrogen scale.

Metals at the top of the table are active or *base*, whilst those at the foot are inactive or *noble.* Thus, in a cell with two dissimilar metals immersed in the same solution, the metal which is higher in this series will tend to dissolve. The electromotive force (e.m.f.) of the cell is the difference of the standard potentials.

A common example of such an electrode system is the Daniell cell (Fig. 3.5). This cell consists of a zinc electrode dipped in dilute sulphuric acid, and a copper electrode in copper sulphate solution contained within a porous pot through which ions can pass from one

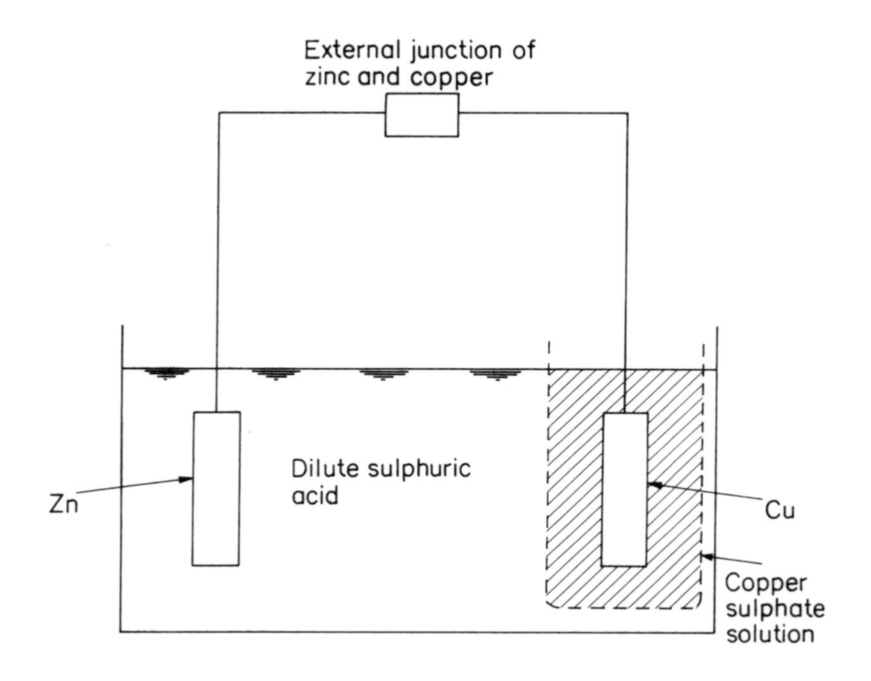

Fig. 3.5 Daniell cell

solution to the other. From Table 3.3, the e.m.f. of the cell is 1·1 V, the difference in the standard potentials between the copper and zinc. At the external junction, electrons move from the zinc to the copper. They thus lower the electron energy levels in the zinc, but since they have a negative charge they raise the positive ion levels in the zinc. Thus, zinc ions will tend to pass into solution. On the other hand, the positive ion levels in the copper will be lowered and copper ions will be deposited on that electrode. At both metal junctions the metal ion moves towards the lower energy state and the energy made available drives current round the circuit. Clearly, the contact potential difference is an important factor in determining the direction of current flow.

Thus at the copper electrode

$$Cu^{++} + 2e \rightarrow Cu$$

and at the zinc electrode

$$Zn \rightarrow Zn^{++} + 2e$$

so that the overall cell reaction is

$$Zn + Cu^{++} \rightarrow Zn^{++} + Cu$$

Note, however, that if a potential difference, greater than the e.m.f. of the cell, is applied in the opposite direction, the current will flow, and the above reactions will take place, in that opposite direction. In ECM, the anode and cathode are usually made of different metals. For instance, suppose that a copper cathode is to be used for the machining of an iron anode. From Table 3.3 we note that the e.m.f. of the cell is 0·78 V, and that the preferred direction for current flow should lead to dissolution of the iron. Nevertheless, since the e.m.f. is so low, the rate of dissolution will be small. The use of an externally applied potential difference across the cell will lead to a greater metal removal rate, and depending upon the direction in which this potential difference is applied, either the iron or the copper will be machined.

3.3.5 Pourbaix diagrams

The electrode potential, E, has been related to the Gibbs free energy, ΔG, by the equation

$$\Delta G = -zEF \qquad (3.13)$$

Now, if a metal is made anodic in an aqueous solution, and if several possible reactions are available, the most likely thermodynamic reaction is that for which the decrease in free energy is greatest. This fact can be explained further by reference to a particular case.

For example, possible reactions for zinc in water are

$$Zn \rightarrow Zn^{++} + 2e \qquad (1)$$

$$Zn + 2H_2O \rightarrow Zn(OH)_2 + 2H^+ + 2e \qquad (2)$$

$$Zn + 2H_2O \rightarrow ZnO_2^{--} + 4H^+ + 2e \qquad (3)$$

$$Zn(OH)_2 + 2H^+ \rightarrow Zn^{++} + 2H_2O \qquad (4)$$

$$Zn(OH)_2 \rightarrow ZnO_2^{--} + 2H^+ \qquad (5)$$

The equilibria (1)–(3) are electrochemical. Reactions (4) and (5) are chemical equilibria. Reaction (1) does not involve hydrogen ions, whilst reactions (2) and (3) involve both hydrogen ions and electrons. The equilibria (4) and (5) are independent of electrons.

We have also seen that, by means of the Nernst equation, the electrode potential at which these reactions occur can be expressed in terms of concentrations, or activities, of the different species involved. For instance, for reaction (2),

$$E = E_0\{\text{Zn}/\text{Zn(OH)}_2\} + \frac{RT}{2F} \ln \frac{[\text{Zn(OH)}_2]\ [\text{H}^+]^2}{[\text{H}_2\text{O}]^2\ [\text{Zn}]}$$

The term $E_0\{\text{Zn}/\text{Zn(OH)}_2\}$ is the normal potential corresponding to reaction (2). As before, the square brackets represent activities. The activities of solid phases and of water present in large excess are usually assumed to be unity, so that we can write the above equation as

$$E = E_0\{\text{Zn}/\text{Zn(OH)}_2\} - \frac{2{\cdot}303RT}{F}\ \text{pH}$$

where $\text{pH} = -\log[\text{H}^+]$. Similar expressions can be derived for the other reactions (1), (3), (4), and (5).

If a value is also chosen for the concentration of the metal ion in solution which corresponds to the condition of no dissolution, (e.g. 10^{-6} mole/l), the linear relationship between the potential and pH can be represented on a 'Pourbaix diagram'.

Since the first of the above five equilibria is independent of pH, the equilibrium potential will be the same for all pH values at which

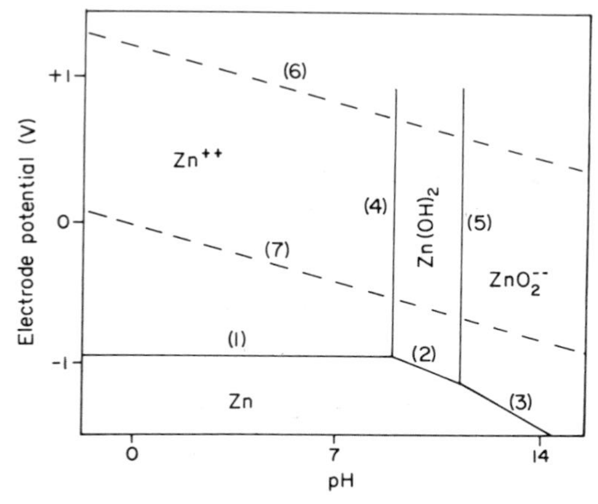

Fig. 3.6 Pourbaix diagram for zinc in water

the reaction is possible. This reaction equilibrium is represented by a horizontal straight line on the Pourbaix diagram for zinc (Fig. 3.6). The equilibria (2) and (3) are dependent on both pH and potential, and are represented by sloping lines. The fourth and fifth equilibria yield vertical lines in Fig. 3.6, since they depend only on pH, and thus the equilibria are the same whatever the potential.

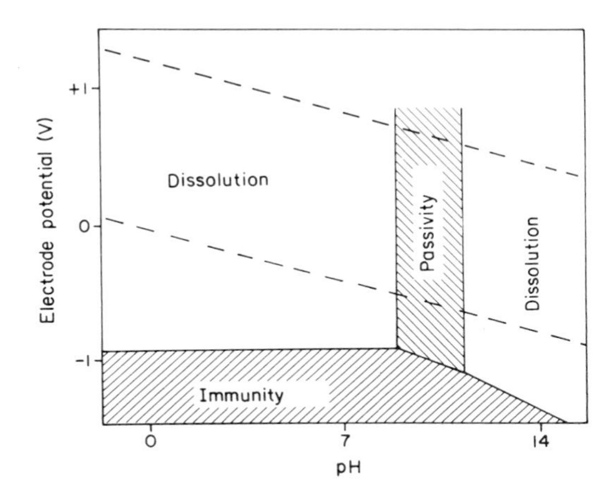

Fig. 3.7 Regions of dissolution, passivity, and immunity for zinc

When the pH of the solution is less than 7, zinc will dissolve at suitable potentials to form freely soluble Zn^{++}. The complex zincate ions ZnO_2^{--} are also forms of zinc in solution, so that the domains in Fig. 3.6 thus marked represent areas where the corrosion of zinc is thermodynamically possible. In solutions where the pH is between 8 and 11, the main product of the anodic reaction is zinc hydroxide. This compound is only slightly soluble in water, and can therefore represent a solid precipitate which may form a protective film on the metal surface. This domain is designated *passivity* because passivation by such a film is thermodynamically possible. The region marked 'Zn' at the foot of the diagram represents solid metal which is thermodynamically stable. That is, dissolution of the zinc is not possible. In this way, the regions of dissolution, passivity, and immunity from dissolution shown in Fig. 3.7 are constructed.

In Fig. 3.6, two other lines are of interest. One is associated with the reaction

$$\tfrac{1}{2}O_2 + 2H^+ + 2e = H_2O \qquad (6)$$

The other line represents

$$2H^+ + 2e = H_2 \tag{7}$$

For potentials greater than those on the line representing (6), water evolves oxygen from an immersed electrode. At potentials less than those given by the slope representing (7), hydrogen is evolved at electrodes dipped in water. The domain between these two lines defines the region of stability of water.

Conditions of pH and potential which lead to passivation are encountered with many metals. For instance, the Pourbaix diagram for titanium (Fig. 3.8) indicates that this metal passivates owing to a surface film of TiO_2. In the same way, oxide formation over a wide range of pH–potential values causes chromium to passivate.

Although potential–pH diagrams have been compiled for all the elements, the information is restricted to reactions between the metal, its oxides, hydroxides, and water [5].

The presence of various anions in ECM electrolytes modifies, and undoubtedly complicates, any simple Pourbaix diagram. Nevertheless, useful information can be obtained. In the case of titanium mentioned above, the indication from the Pourbaix diagram that

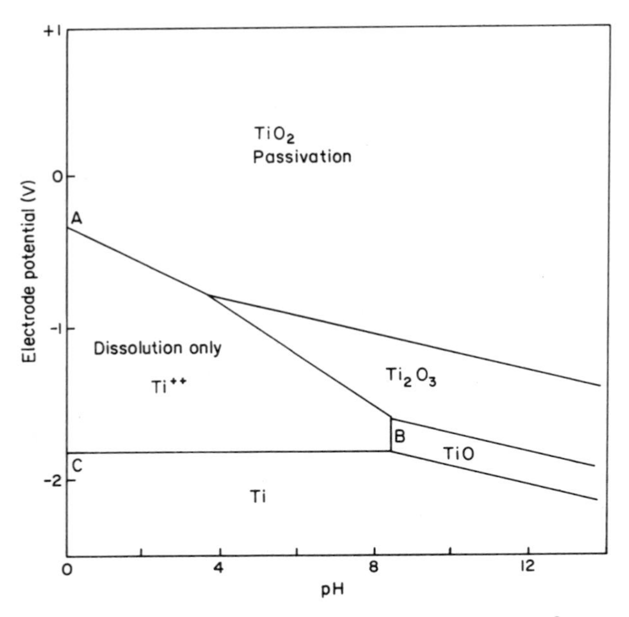

Fig. 3.8 Pourbaix diagram for titanium at 25°C

passivation is likely due to a surface film means that appropriate action can be taken by the addition to the electrolyte of a complexing agent which will dissolve the oxide film.

Even less information is available about the application of Pourbaix diagrams to alloys. Again, though, some deductions from the behaviour of elements can be made. The passivating nature of chromium is known to contribute substantially to the passivity of stainless steel, because of its large chromium content. In like fashion, behaviour about titanium alloys can be established from that of titanium metal.

The anodic processes considered so far in this section refer, of course, only to conditions of thermodynamic equilibrium. The non-equilibrium processes at work in ECM can be understood from the kinetics of dissolution.

3.4 Irreversible electrode reactions

In ECM, we have observed that metal dissolution is achieved by the deliberate application of an external potential difference which gives rise to a corresponding current flow. It will be shown below that, the greater the current flow, the greater is the difference in potential between the equilibrium value and the *working* value. The electrode reaction is now said to be *irreversible*, and the difference between the equilibrium and working values of the potentials is known as the *overpotential.* It is commonly accepted that, in most electrochemical reactions, there are three specific types of over-potential: activation, concentration, and resistance. Each type is now discussed.

3.4.1 Activation overpotential

Suppose a potential difference is now applied across the cell to cause anodic dissolution. The anode electrode must then ionise at a greater rate than that of discharge of its ions. The electrode potential is accordingly altered from its equilibrium value by an amount η_a, the *activation overpotential*, and the energy–distance diagram of Fig. 3.9 now replaces the equilibrium diagram of Fig. 3.3. The free energy required for dissolution is now reduced from ΔG_1 to $(\Delta G_1 - z\alpha\eta_a F)$, and the energy for discharge is increased from ΔG_2 to $[\Delta G_2 + z(1 - \alpha)\eta F]$ where α is that fraction of overpotential associated with dissolution.

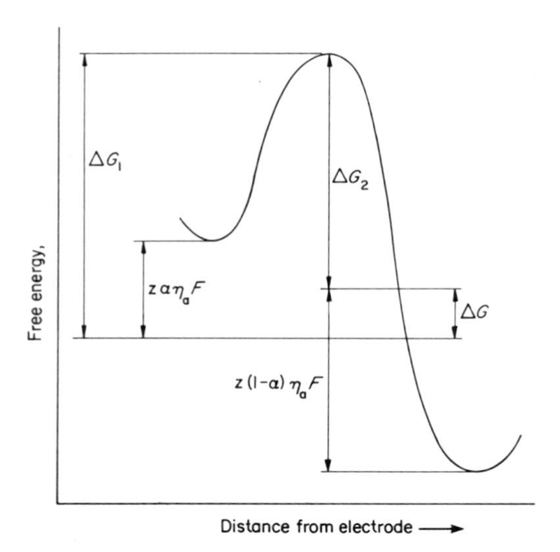

Fig. 3.9 Free energy–distance diagram for irreversible conditions of anodic dissolution

Under these conditions, the current density J related to the dissolution becomes

$$J = \Omega_1 \exp\left[-\frac{(\Delta G_1 - z\alpha\eta_a F)}{RT}\right] - \Omega_2 \exp\left[-\frac{(\Delta G_2 + z(1-\alpha)\eta_a F)}{RT}\right]$$

$$= J_0 \left[\exp\left(\frac{z\alpha\eta_a F}{RT}\right) - \exp\left(\frac{z(1-\alpha)\eta_a F}{RT}\right)\right] \qquad (3.14)$$

3.4.2 *The Tafel equation*

It will be readily appreciated that ECM is a highly irreversible process, so that the rate of reaction in the direction opposed by the overpotential is negligible. Numerically, this means that η_a probably exceeds 0·05 V and that the second term on the right-hand side of Equation (3.14) may be ignored. The equation then becomes

$$\eta_a = \left(\frac{2{\cdot}303RT}{z\alpha F}\right) \log J/J_0$$

$$= a + b \log J \qquad (3.15)$$

where

$$a = -\frac{2{\cdot}303RT}{z\alpha F} \log J_0; \quad b = \frac{2{\cdot}303RT}{z\alpha F}$$

Table 3.4 Overpotential values for hydrogen evolution in 1N H_2SO_4 (current density 10^{-2} A/cm^2) (after Evans [7])

Cathode material	Overpotential (V)
Pt, bright	0·07
Fe	0·56
Ni	0·74
Cu	0·58

are the Tafel constants. A similar expression can be derived for the cathodic reaction.

At this stage it is worth stressing the extreme nature of the electrode processes at work in ECM. The large currents involved and steep concentration gradients (see below) in the immediate vicinity of the metal–solution interface may induce a more complex form for the activation overpotential.

Tafel overpotentials have been studied mainly in connection with hydrogen evolution at the cathode although some information has been collected on anodic phenomena [6]. Values are found to depend on a number of factors including the condition and material of the electrode. Some data [7] for overpotentials for hydrogen evolution from 1N H_2SO_4 are available in Table 3.4. Further values for the Tafel constants are presented in Table 3.5.

The current densities for which these results have been obtained are much lower than those used in ECM. However, an investigation has been reported [8] in which hydrogen overpotentials at current densities up to 10^2 A/cm^2 have been measured, other effects such

Table 3.5 Values of the Tafel constants a and b, at 20°C (after Evans [7])

Cathode material	Electrolyte solution		a (V)	b (V)
Fe	1N	HCl	0·70	0·125
	2N	NaOH	0·76	0·112
Ni	0·11N	NaOH	0·64	0·100
Cu	1N	H_2SO_4	0·80	0·115

as heating and concentration overpotentials, having been made negligible by high solution velocities (above 20 m/s). In particular from those results, for a nickel cathode in 1N HCl, the Tafel relationship was found to hold only for current densities up to 1 A/cm^2. Above that value the overpotential increased substantially with current density (e.g. about 1 V at 10 A/cm^2). The need for information relevant to ECM has led to studies of anode potentials in high-rate anodic dissolution of copper [9]. Tafel behaviour was observed up to 39 A/cm^2 for copper dissolved in 1M H_2SO_4 + 0·1M $CuSO_4$ solution, the maximum electrolyte flow-rate being 10 m/s.

The accuracy of the results is also discussed in that work; the two main factors which influence accuracy are the *IR* drop in the solution and effects due to the double layer. It is also relevant to mention some observations by Mao [10]. He reports that no hydrogen is formed at a brass cathode during ECM of mild steel in 2 to 4·5M $NaNO_3$ at current densities ranging from 12 to 47 A/cm^2. But Mao did find that a small amount of hydrogen is generated at a platinum electrode in the same electrolyte because of the comparatively small hydrogen overpotential on platinum. This characteristic low overpotential is the reason for the wide use of platinised platinum electrodes in studies of hydrogen evolution.

Example

A useful estimate of the value of Tafel overpotential for ECM conditions can be obtained from Equation (3.15). For typical values, $RT/zF = 1/40$ V, $\alpha \simeq \frac{1}{2}$, $J_0 \simeq 10^{-5}$ A/cm^2, $J = 10^2$ A/cm^2, the Tafel constants a and b are calculated to be 0·58 V and 0·12 V respectively, and η_a is found to be 0·82 V. This is a significant amount in comparison with a typical applied potential difference of 10 V.

3.4.3 Concentration overpotential

As dissolution proceeds in ECM, the movement of ions is controlled by three processes:

(i) migration, i.e. movement under the influence of the electric field;

(ii) convection, i.e. bodily movement of the electrolyte solution; in ECM this is mainly effected by forced agitation of the electrolyte;

(iii) diffusion, i.e. movement due to ion concentration gradients in the solution.

In studies of these processes, the concept of the diffusion layer is very useful. Thus, consider the anodic dissolution reaction. When the rate of metal dissolution is greater than the rate at which the metal ions can diffuse away from an electrode, a condition is reached in which an ionic concentration gradient exists over a thin layer of electrolyte adjacent to the electrode. This layer is called the 'diffusion layer'. A change in the electrode potential from the reversible value occurs due to this concentration gradient, and the numerical difference between the reversible and new value is said to be the *concentration overpotential*, η_{conc}. We shall now estimate the magnitude of η_{conc} [11].

Usually the outer edge of the diffusion layer is assumed to be plane and parallel to the electrode surface. Its thickness is designated by δ. Next, consider a volume element within the diffusion layer, and the net rate of flow of particles entering and leaving it. The number of particles per second passing through the surface S surrounding the volume V is $-\oint j\, dS$, where j is the number flux (i.e. the number of particles per second passing through unit area of the surface). Suppose that c is the concentration, i.e. the number of particles per unit volume of solution for each ion. (Later, c will be related to C, the bulk concentration of the electrolyte.) Then if $\partial c/\partial t$ is the change in the number of particles per unit volume per second, the change in the number of particles in the volume V is $\int(\partial c/\partial t)\mathrm{d}V$. Hence

$$\int \frac{\partial c}{\partial t}\,\mathrm{d}V = -\oint j\,\mathrm{d}S$$

Using the Gauss theorem, we deduce that

$$\int \frac{\partial c}{\partial t}\,\mathrm{d}V = -\int \mathrm{div}\, j\,\mathrm{d}V$$

or

$$\frac{\partial c}{\partial t} + \mathrm{div}\, j = 0 \qquad (3.16)$$

The number flux j receives contributions from the diffusion flux ($= -D$ grad c) due to the concentration gradient, the migration flux ($= zeE_\delta\, uc$)due to the electric field, and the convection flux ($= c\boldsymbol{U}$) due to the movement of the ions by the solution flow. D is the diffusion coefficient; although it is dependent upon concentration and temperature, it is assumed constant here. $\boldsymbol{E}_\delta$ is the electric field strength across the diffusion layer, u the ionic mobility, e the electronic charge, and U the local solution velocity. We then have

$$\frac{\partial c}{\partial t} = -\text{div}(-D \text{ grad } c + zeE_\delta uc + c\boldsymbol{U})$$

Since D is constant, div(D grad c) = D div grad c. Also, div($c\boldsymbol{U}$) = ($\boldsymbol{U}$ grad)c + c div $\boldsymbol{U}$. The electrolyte solution being assumed to be incompressible, div $\boldsymbol{U}$ = 0. Finally, we consider that the thickness of the diffusion layer is very small compared with the gap width, so that transport of ions by convection within it can be neglected. Then we can assume that $\boldsymbol{U}$ = 0. (Note, however, that the solution velocity is relevant in the determination of the thickness of the diffusion layer.) The position of each volume element of the solution is now specified by its distance y from the electrode surface. If the number of ion types is restricted to metal, positive and negative electrolyte ions (assuming that the electrolyte is a simple typical one like NaCl), and if the respective subscripts m, +, and − are used, Equation (3.16) becomes, for each type

$$\begin{aligned}
\frac{\partial c_{\mathrm{m}}}{\partial t} + \frac{\partial}{\partial y}\left(-D_{\mathrm{m}} \frac{\partial c_{\mathrm{m}}}{\partial y} + z_{\mathrm{m}} eE_\delta u_{\mathrm{m}} c_{\mathrm{m}}\right) &= 0 \\
\frac{\partial c_+}{\partial t} + \frac{\partial}{\partial y}\left(-D_+ \frac{\partial c_+}{\partial y} + z_+ eE_\delta u_+ c_+\right) &= 0 \qquad (3.17) \\
\frac{\partial c_-}{\partial t} + \frac{\partial}{\partial y}\left(-D_- \frac{\partial c_-}{\partial y} - z_- eE_\delta u_- c_-\right) &= 0
\end{aligned}$$

Einstein's relation between diffusion coefficient D and mobility u for each ion type is also useful:

$$D = ukT \qquad (3.18)$$

In the steady-state, time-independence of the concentrations and integration of Equation (3.17) give

$$\frac{-\mathrm{d}c_{\mathrm{m}}}{\mathrm{d}y} + \frac{z_{\mathrm{m}} eE_\delta c_{\mathrm{m}}}{kT} = A \qquad (3.19)$$

$$\frac{-\mathrm{d}c_+}{\mathrm{d}y} + \frac{z_+ eE_\delta c_+}{kT} = 0 \qquad (3.20)$$

$$\frac{-\mathrm{d}c_-}{\mathrm{d}y} - \frac{z_- eE_\delta c_-}{kT} = 0 \qquad (3.21)$$

where A is the constant of integration which is to be determined. If electrical neutrality within the diffusion layer is assumed, then

$$z_m c_m + z_+ c_+ - z_- c_- = 0 \tag{3.22}$$

For simplicity, suppose that

$$z_m = z_+ = z_- = 1$$

Equation (3.22) becomes

$$c_m + c_+ = c_- = \tfrac{1}{2}c, \text{ say} \tag{3.23}$$

Addition of the three equations (3.19), (3.20), and (3.21) and the use of Equation (3.23) yield, on integration,

$$c = -Ay + B \tag{3.24}$$

where B is a constant. At $y = 0$, let $c = c_0$, and at $y = \delta$ let $c = c_\delta$. Then

$$A = \frac{c_0 - c_\delta}{\delta}$$

The current density across the diffusion layer is now given by

$$\begin{aligned} J &= eD_m A \\ &= eD_m \left(\frac{c_0 - c_\delta}{\delta}\right) \end{aligned} \tag{3.25}$$

Now, since $E_\delta = -\mathrm{d}\phi/\mathrm{d}y$ in the layer, from Equations (3.21) and (3.23),

$$\mathrm{d}\phi = \frac{kT}{e}\frac{\mathrm{d}c}{c}$$

which, on integration, gives

$$\eta_{conc} = \frac{kT}{e} \ln \frac{c_\delta}{c_0} \tag{3.26}$$

Substituting from Equation (3.25) for c_0, we obtain

$$\eta_{conc} = -\frac{kT}{e} \ln\left(1 + \frac{\delta J}{eD_m c_\delta}\right) \tag{3.27}$$

The electrolyte is assumed to be sufficiently agitated so that c_δ takes the value of the bulk concentration of the electrolyte, C. The thickness of δ also depends on the electrolyte flow conditions. (In this section a subscript m has had to be used to distinguish the properties of the metal ions from those of the other (electrolyte) ions. The symbol used elsewhere for, say, metal ion valency, $z+$ is identical with z_m).

3.4.4 *Thickness of the diffusion layer*

In the determination of δ, three flow conditions for the electrolyte need consideration.

For unstirred electrolytes, δ is usually found to be about 10^{-2} mm. (For instance, Higgins [12] calculates that δ is $2{\cdot}8 \times 10^{-2}$ mm for nickel dissolving at $2{\cdot}54$ A/cm^2 in 1N HCl.)

Several methods have been put forward from which the thickness of δ for flowing conditions can be estimated.

One analysis [13] has been based on the equations for convective diffusion (3.17) which, as noted, describe the transport of dissolved particles in terms of the concentration gradients and the components of the solution velocity. There is great similarity between these equations and the Navier–Stokes equations. They can be reduced, therefore, to describe the behaviour of the solution over the diffusion layer in the same way that Prandtl's boundary layer equations have been derived. By an order of magnitude analysis, similar to that used in Chapter 2 to obtain an estimate of the boundary layer thickness, δ can be deduced to be given by

$$\delta = \delta_0 \left(\frac{D}{\nu}\right)^{1/3} \tag{3.28}$$

where δ_0 is the hydrodynamic, laminar boundary layer thickness, D is the diffusion coefficient and ν the kinematic viscosity of the solution.

Since $D \simeq 10^{-5}$ cm^2/s, $\nu \simeq 1$ mm^2/s, δ is about one tenth of the thickness of δ_0. This result can be related to the expression derived in Chapter 2 for δ_0:

$$\delta_0 = 5\left(\frac{\nu x}{U}\right)^{1/2}$$

where U is the mainstream velocity.

Note that the thickness of the diffusion layer increases in the downstream direction, and decreases with increasing velocity. This theory has been utilised [14] to obtain an expression for the average thickness of the diffusion layer $\overline{\delta}$ over an electrode length L:

$$(\overline{\delta})^{-1} = \left(\frac{2}{3}\right)\left(\frac{\nu}{D}\right)^{1/3}\left(\frac{U}{\nu L}\right)^{1/2} \tag{3.29}$$

(As pointed out in Chapter 2, in this work a numerical coefficient of three, rather than the more usual five, is used in the Prandtl expression for the laminar boundary layer thickness.) In a related experimental study, in which $L = 2$ mm, $h = 0{\cdot}1$ mm, $U \simeq 9$ m/s (calculated from volume flow-rate at inlet), the average diffusion layer thickness was found to be about $1{\cdot}5 \times 10^{-3}$ mm. The theoretical comparison from Equation (3.29) gives $2{\cdot}2 \times 10^{-3}$ mm, with the assumed values $D = 10^{-5}$ cm²/s and $\nu = 1$ mm²/s.

In turbulent flow the hydrodynamic viscous sub-layer has its analogous diffusion sub-layer. The thicknesses of these layers have been related by the expression

$$\delta = \left(\frac{D_m \delta_1^3}{\gamma u_f}\right)^{1/4} \qquad (3.30)$$

where δ_1 is the thickness of the viscous sub-layer, u_f is the friction velocity, and γ is a dimensionless constant, found experimentally to be approximately unity.

From the previous chapter, Equation (2.42) gives

$$\delta_1 \simeq \frac{5\nu}{u_f}$$

so that on substitution into Equation (3.30)

$$\delta = \delta_1 \left(\frac{D_m}{5\nu}\right)^{1/4} \qquad (3.31)$$

It was also shown in Chapter 2 that δ_1 can be obtained from the solution of the equation

$$\delta_1 = \frac{5\nu C_1}{U} \ln\left(\frac{\delta_1}{2h}\right)$$

where $C_1 \simeq -2.5$. From this equation and Equation (3.31), δ can be calculated.

Example

From experimental results [15], in which $h = 0{\cdot}43$ mm, $U = 9{\cdot}2$ m/s, with $\nu = 1$ mm²/s and $D_m = 10^{-5}$ cm²/s (assumed), δ_1 is calculated to be $6{\cdot}6 \times 10^{-3}$ mm. From Equation (3.31) the corresponding thickness of the diffusion layer thickness δ is $7{\cdot}9 \times 10^{-4}$ mm.

An alternative method for estimating the average diffusion layer thickness is available [16, 17]. Here δ is related to the average Nusselt number, Nu, by

$$\delta = \frac{d_h}{\mathrm{Nu}} \tag{3.32}$$

where d_h is the hydraulic diameter for the flow channel. For fully developed velocity profiles, and at short distances from the leading edge of the electrode, experimental mass transfer rates are described by the relation

$$\mathrm{Nu} = 0{\cdot}28\ \mathrm{Re}^{0{\cdot}58}\ \mathrm{Sc}^{1/3} \left(\frac{d_h}{L}\right)^{1/3} \tag{3.33}$$

where Nu is defined by

$$\mathrm{Nu} = \frac{J_l d_h}{zFC_b D} \tag{3.34}$$

Here J_l is the limiting current density, z is the number of electrons transferred per mole, F is Faraday's constant, C_b is the bulk concentration of the reacting species. Re is the Reynolds number, and Sc ($= \nu/D$) is the Schmidt number, ν and D being the viscosity and diffusion coefficient respectively.

Relation (3.33) applies if the condition

$$\frac{0{\cdot}2L\ \mathrm{Re}^{7/8}}{d_h} < 10^3 \tag{3.35}$$

is satisfied.

If

$$\frac{0{\cdot}2L\ \mathrm{Re}^{7/8}}{d_h} \gg 10^3 \tag{3.36}$$

the mass transfer rates are given by

$$\mathrm{Nu} = 0{\cdot}022\ \mathrm{Re}^{7/8}\ \mathrm{Sc}^{1/4} \tag{3.37}$$

(Note: expression (3.37) is derived [16] with the aid of a friction factor which is made equal to $0{\cdot}079/\mathrm{Re}^{1/4}$).

For laminar flow, the equivalent expression for fully developed velocity profiles is

$$\mathrm{Nu} = 1{\cdot}85 \left(\mathrm{Re}\ \mathrm{Sc}\ \frac{d_h}{L}\right)^{1/3} \tag{3.38}$$

Substitution of typical values $\nu = 1\ mm^2/s$, $D = 10^{-5}\ cm^2/s$, $h = 0{\cdot}5$ mm ($d_h = 2h$) into these formulae has shown that over the mass transfer entry region, δ is about 4×10^{-3} to 8×10^{-4} mm for velocities ranging from 1 to 25 m/s respectively [16].

Example

The magnitude of the concentration overpotential η_{conc} can now be estimated from Equation (3.27), the appropriate values of δ for the different flow conditions being obtained from the relevant formulae above.

Suppose $J = 40\ A/cm^2$, $c_\delta = 1{\cdot}7$ N, $kT/e = 1/40$ V, $D_m \simeq 10^{-5}\ cm^2/s$, $\nu \simeq 1\ mm^2/s$, $\delta \simeq 1{\cdot}5 \times 10^{-3}$ mm (laminar flow), and $\delta \simeq 7{\cdot}9 \times 10^{-4}$ mm (turbulent flow). $|\eta_{conc}|$ is calculated to be 0·027 V and 0·038 V respectively for the smaller and larger values of δ. Both values for the overpotential are considerably less than a typical applied potential difference of 10 V. It can be inferred that, under the usual conditions of electrolyte flow in ECM, concentration polarisation is negligible.

Fitz-Gerald and McGeough [18] have also postulated that convection in the diffusion layer may cause considerable variation in the diffusion currents along the electrode, thus causing a supplementary overpotential effect which is dependent on the electrolyte velocity. They have called this possible effect the 'convection overpotential'.

In their analysis, the transport equations for the metal and electrolyte ions are modified to include a term for convection of ions within the diffusion layer. The presence of this term leads to expressions for the concentration gradient which are dependent on the electrolyte flow velocity:

$$\frac{\partial c}{\partial y} \propto x^{-(1-s)/3}$$

where $s = -\frac{1}{2}$ for laminar flow and $s = 0$ for turbulent flow. The current density at the anode, of course, has a similar dependence; therefore,

$$J \propto x^{-1/2}$$

for laminar flow, and

$$J \propto x^{-1/3}$$

for turbulent flow.

Example

For turbulent flow and for an electrode of length 80 mm, say, in the absence of other effects, the current density at $x = 80$ mm is one-half of that at $x = 10$ mm.

Only steady-state conditions have been considered in the analyses in this section. For electrode processes under diffusion control the transition time required for equilibrium to be reached has been shown [19] to be estimable from the expression

$$Jt^{1/2} = \text{constant}$$

For electrodeposition processes, $t \simeq 0{\cdot}1$ to 1 s. For ECM, t will probably be less than 10^{-4} s. For conditions in which a limiting concentration of reaction products is obtained at the electrode, passivation (see below) might occur in a shorter time.

In relation to this section, it is interesting finally to note that an experimental technique has been developed for analysing the properties of the anode diffusion layer, including its thickness, for conditions akin to those met in ECM [20].

3.4.5 **Resistance overpotential, η_r**

Resistance overpotential is generally regarded as the potential drop across a thin layer of electrolyte or film layer (e.g. an oxide film) on the electrode surface. Its magnitude depends principally on the current flowing in the cell and on the nature and conductivity of the electrolyte. For a small volume element of the electrolyte, the resistance is inversely proportional to the concentration, and by Ohm's law, the current at which η_r becomes appreciable is proportional to the electrolyte concentration.

There is clearly a connection between the resistance and activation overpotentials. In practical cases, the presence of the former is usually noticed by the need to increase the voltage required to maintain the reaction.

The principal differences between η_a and η_r are in their different variations with current density and in their rates of build-up and decay; η_r forms or decays almost instantaneously while η_a forms and decays quickly but measurably.

3.4.6 **Decomposition potential**

Generally, in ECM, the higher the applied potential difference the greater is the rate of metal dissolution at the anode (and hydrogen evolution at the cathode). The characteristic shape of the curve of applied potential difference against current is shown in Fig. 3.10. For the low, initial values of potential difference the current is low

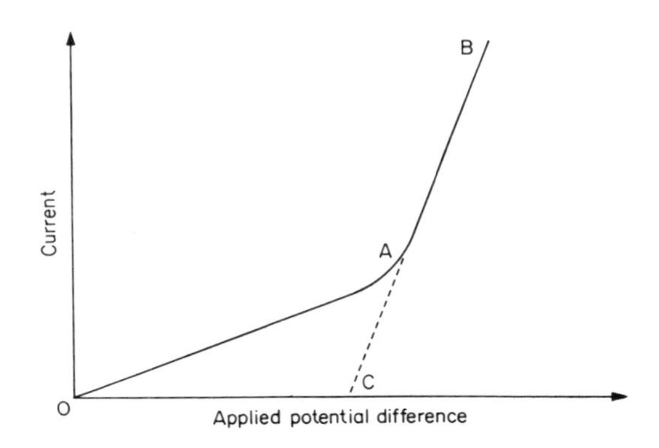

Fig. 3.10 Decomposition potential

(along OA) but near values of the potential difference corresponding to region A, the current rises sharply. If the potential difference is further increased, the cell current still increases appreciably (along AB). That part of the curve corresponding to A represents the onset of anodic dissolution and cathodic gas evolution of the process. If the section AB is extrapolated back to zero current, the value of the potential at that point C is known as the *decomposition potential.*

A typical example for ECM is the electrolysis involving the dissolution of iron in NaCl solution, discussed in Chapter 1. At the cathode, the molecular hydrogen will only be formed when the reversible hydrogen electrode potential is exceeded. Similarly, the dissolution of iron will only occur when its reversible electrode potential is exceeded. Thus, the potential difference between the electrodes which must be exceeded so that the overall cell reaction can take place is given by the difference between the reversible potentials for the hydrogen and iron. If the hydrogen electrode potential is taken as zero, then from Table 3.3 the minimum potential difference required for anodic dissolution is seen to be the same as the reversible potential for iron (0·44 V).

However, it has been observed that the actual potential difference in ECM is greater than this minimum potential difference, because of the irreversibility of the process (and the consequent presence of overpotentials). Thus, to maintain a current I in the cell the applied potential difference V necessary is given by

$$V = (E_c - E_a) + \eta_a + \eta_{conc} + \eta_r + IR$$

where E_c and E_a are the reversible potentials at the cathode and anode respectively, η_a, η_{conc}, and η_r are the overpotentials at both electrodes, as discussed above, and IR is the ohmic potential drop across the electrode gap, R being the resistance of the electrolyte.

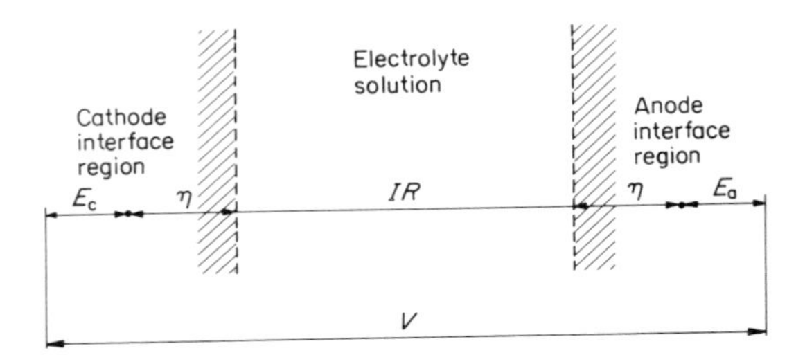

Fig. 3.11 Schematic potential distribution in an ECM cell

In Chapter 1 the IR drop was shown to be dependent on the cell conditions of current density, gap width, and electrolyte conductivity ($IR \equiv Jh/\kappa_e$). In practice η_{conc} and η_r are often small compared with η_a, so that, in calculations, only the activation overpotentials at anode and cathode need consideration. Note, however, that an additional contribution is possible from the voltage drop across the leads. A very simple schematic representation of the potential distribution for an ECM cell is given in Fig. 3.11.

It is worth mentioning at this stage the possibility of anodic reactions other than, or in addition to, dissolution. In aqueous solutions, few metals dissolve without the generation of gas (e.g. oxygen) at sufficiently high potential differences. Since part of the current is now used in evolving the gas, a practical observation then is the decrease in the metal dissolution rate below the Faraday rate. Gas evolution is discussed more fully in the next section.

3.5 Polarisation curves

During anodic dissolution, a curve of anodic potential against current density can be obtained. The usefulness of such a polarisation curve will become clear, and will be discussed more fully, in Chapter 4. However, since its form relates to some of the anodic phenomena discussed above, its main features are outlined here. A simple, characteristic polarisation curve is shown in Fig. 3.12. For anodic potentials in the range AB, the metal is said to be in the 'active' state, and dissolves by the removal of cations from the crystal planes.

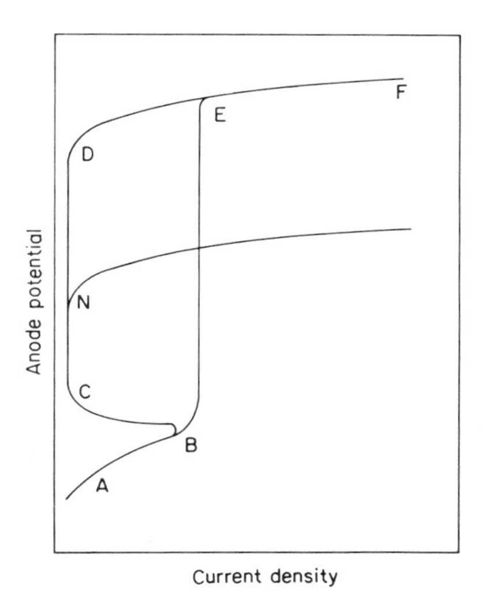

Fig. 3.12 Simple illustration of polarisation curve

Since the rates of dissolution depend on the geometry of the plane, the metal surface becomes etched. Under these conditions, Tafel's equation, derived earlier, applies to the current density–potential relationship.

When the anode potential becomes greater than B in Fig. 3.12, a solid oxide film may form on the anode, which causes the reaction rate to be reduced, for example, to the level CD. The metal is then said to have become 'passive'. The solid films associated with this condition are often similar to those already discussed in relation to the thermodynamics of equilibrium anodic processes. (In passing we note that the passivity of iron in concentrated HNO_3 was first observed by Faraday.)

Alternatively, surface films may be formed on metals which are so conductive that the passage of a considerably greater amount of current becomes possible. This condition is shown as the region BE in Fig. 3.12. When the oxide films are present in this state, dissolution takes place in a random manner over the metal surface. (It is now controlled by the film and not by the geometry of the crystal plane.) Since the atoms are removed in a random fashion, the metal surface becomes polished.

For oxide films with good electronic conductivity, and for appropriate conditions of potential, the oxidation of solution anions may take place in preference to, or in addition to, polishing or passivity. This reaction yields the evolution of gas; e.g. for hydroxyl ions, oxygen is formed:

$$4OH^- \rightarrow 2H_2O + O_2 + 4e$$

This condition is represented by DF in Fig. 3.12.

A condition which is sometimes difficult to distinguish from gas evolution is transpassivity. This condition arises when cations of the oxide film are themselves oxidised to soluble higher-valency forms. Chromium is one metal which is known to exhibit this condition yielding chromates.

If a passive film is present on a metal and the metal is exposed to a solution containing an appropriate concentration of aggressive anions (e.g. chloride ions), those anions will penetrate into the film, causing a disruption of the film. (This condition is well known in corrosion studies.) The areas of penetration permit greater local current flow than elsewhere over the electrode where the film is unbroken. Such local current flow can lead to the formation of pitting on the electrode surface. Pitting is most likely to occur at weak points in the oxide film, e.g. at grain boundaries. The potential at which the oxide film is disrupted is the 'critical breakdown potential' and is dependent on the concentration of the aggressive ion. The effect of increasing chloride concentration on the potential which gives rise to pitting is given by D and N in Fig. 3.12. The increase in current after pitting is due to the exposure of the metal surface within the pits to the electrolyte.

Several final remarks must be made about polarisation curves. They are usually derived from experiments carried out at low or zero electrolyte velocities and low current densities. An increase in electrolyte velocity causes little change in the shape of the polarisation curve apart from increasing any limiting value of the current density. This is demonstrated in Fig. 3.13 for an electropolishing curve. This figure is also presented in an alternative form – with current density and potential axes interchanged – which is also popular in polarisation work. Little work has been done to study the anodic polarisation characteristics of the alloys used in ECM.

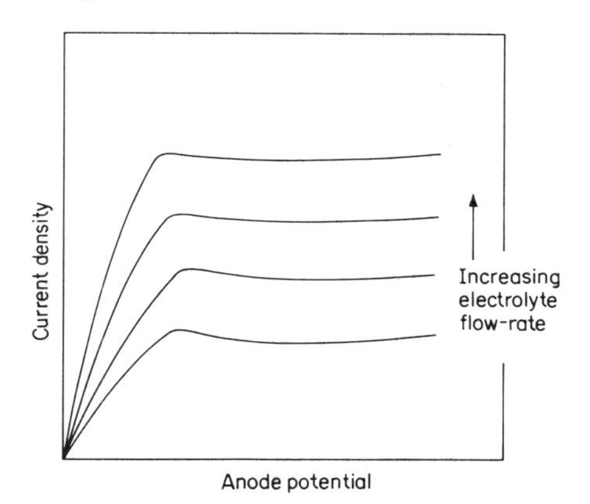

Fig. 3.13 Increase in limiting current density with electrolyte flow-rate

Nonetheless, a start has been made in reported studies of the behaviour of copper, zinc, and brass [21] and iron–chromium alloys [22].

3.6 Reactions at the electrodes

Common reactions which occur in ECM for different types of electrolytes are finally summarised.

(*a*) *Acidic electrolytes* (e.g. HCl)

Anodic reaction: dissolution of the metal:

$$M = M^{z+} + ze$$

where M^{z+} is the metal ion of valency z, and e is the electron charge.
Cathodic reaction: evolution of hydrogen:

$$2H^{+} + 2e = H_2$$

Reduction of the positive metal ion is also possible, but is undesired.

$$M^{z+} + ze = M$$

(*b*) *Neutral electrolytes* (e.g. NaCl)

Anodic reaction: metal dissolution

$$M = M^{z+} + ze$$

or

$$M + z(OH)^{-} = M(OH)_z + ze$$

Cathodic reaction: hydrogen generation

$$2H_2O + 2e = H_2 + 2OH^-$$

The metal ions from the anode and hydroxyl ions from the cathode usually react in the bulk of the electrolyte to form a metal hydroxide:

e.g. $$M^{2+} + 2(OH)^- = M(OH)_2$$

(*c*) *Alkaline electrolytes*

Anodic reaction: metal dissolution

$$M + z(OH)^- = M(OH)_z + ze$$

Cathodic reaction: hydrogen generation

$$2H_2O + 2e = H_2 + 2OH^-$$

References

1. *International Critical Tables*, McGraw-Hill, New York (1933).
2. Kaye, G. W. C. and Laby, T. H., *Tables of Physical and Chemical Constants*, Longmans, London (1962).
3. Butler, J. A. V., *Electrocapillarity*, Methuen, London (1940). Ch. 3.
4. See, for example, Gurney, R. W., *Ionic Processes in Solution*, Dover, New York (1962) Ch. 6.
5. Pourbaix, M., *Atlas d'Equilibres Electrochemiques*, Gauthier-Villars, Paris (1963).
6. Hoar, T. P., in *Modern Aspects of Electrochemistry* (ed. J. O'M. Bockris), Butterworths, London, No. 2, p. 274.
7. Evans, U. R., *Introduction to Metallic Corrosion*, Arnold, London (1963) p. 1025.
8. Azzam, A. N. and Bockris, J. O'M., *Nature* (1950) **165**, March 11, 403.
9. Landolt, D., Muller, R. H. and Tobias, C. W., *J. Electrochem. Soc.* (1971) **118**, No. 1, 40.
10. Mao, K. W., *J. Electrochem. Soc.* (1971) **118**, No. 11, 1876.
11. Fitz-Gerald, J. M. and McGeough, J. A., *J. Inst. Maths. Applics.* (1969) **5**, 387.
12. Higgins, J. K., *J. Electrochem. Soc.* (1959) **106**, 999.
13. Levich, V. G., *Physicochemical Hydrodynamics*, Prentice-Hall, New York (1962) Ch. II, III.
14. Kruissink, C. A., Paper presented at First Int. Conf. on ECM, Leicester University, March 1973,
15. Hopenfeld, J. and Cole, R. R., *Trans. ASME, J. Eng. Ind.* (1966) 455.
16. Landolt, D., Muller, R. H. and Tobias, C. W., *J. Electrochem. Soc.* (1969) **116**, No. 10, 1384.
17. Landolt, D., Muller, R. H. and Tobias, C. W., in *Fundamentals of ECM* (ed. C. L. Faust), Electrochemical Society Softbound Symposium Series, Princeton (1971) p. 200.

18. Fitz-Gerald, J. M. and McGeough, J. A., I.E.E. Conf. on Electrical Methods of Machining, Forming and Coating, London, 1970, Conf. Publication No. 63, p. 72.
19. West, J. M., *Electrodeposition and Corrosion Processes*, Van Nostrand-Reinhold, London, 2nd edition (1970).
20. Flatt, R. K., Wood, J. W. and Brook, P. A., *J. Appl. Electrochem.* (1971) **1**, 35.
21. Flatt, R. K. and Brook, P. A., *Corrosion Science* (1971) **11**, 185.
22. Simpson, J. P. and Brook, P. A., Paper presented at First Int. Conf. on ECM, Leicester University, March 1973.

Bibliography

Bockris, J. O'M. and Drazic, D., *Electrochemical Science*, Taylor and Francis, London (1972).
Butler, J. A. V., *Electrocapillarity*, Methuen, London (1940) Ch. 3.
Gurney, R. W., *Ionic Processes in Solution*, Dover, New York (1962).
Levich, V. G., *Physicochemical Hydrodynamics*, Prentice-Hall, New York (1962).
Potter, E. C., *Electrochemistry, Principles and Applications*, Cleaver-Hume, London (1961).
Stewart, D. and Tulloch, D. S., *Principles of Corrosion and Protection*, Macmillan, London (1968).
Vetter, K. J., *Electrochemical Kinetics*, Academic Press, London (1967).
West, J. M., *Electrodeposition and Corrosion Processes*, Van Nostrand-Reinhold, London, 2nd edition (1970).

CHAPTER FOUR

Metals and Electrolytes in ECM

ECM is unlike other, well established anodic dissolution processes in that the current density is high, and the electrolyte is in motion at a high velocity, and is highly concentrated. These features of the process mean that most electrochemical data, which are concerned with low current densities, and with electrolytes which are unstirred or slowly moving, and of low concentration, cannot necessarily be related to

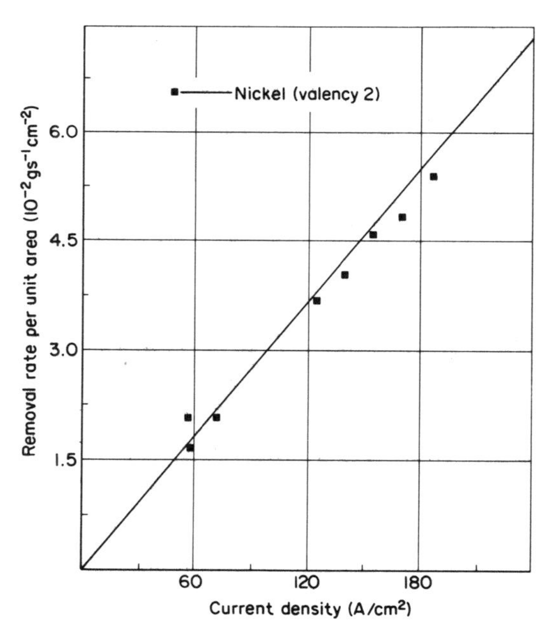

Fig. 4.1 Removal rate as a function of average current density for nickel

it. As a result, since the advent of ECM, much attention has been paid to the rates and modes of reactions at both electrodes for conditions comparable to those met in ECM; for not only must the rate at which metal is removed from the anode be sufficiently high, but also the manner of dissolution must leave an acceptable surface finish. Moreover, the progress at the anode should not receive deleterious interference from the complementary reactions at the cathode.

Although the rate of removal in theory is governed only by the local current density, in practice this rate is strongly influenced by other factors, which include the process variables and the relationship between the metal and the electrolyte. The mode of removal is similarly affected. In particular, the metal/electrolyte relationship is closely linked with the formation of surface films which play a dominant part in determining both the rate and the manner of dissolution, and with the properties of those films lies the key to good dimensional control as well as to the quality of surface finish in ECM.

4.1 Theoretical removal rates for elements

The procedure for calculating removal rates for elements has been stated in Chapter 1. If the atomic weight A and valency z of the dissolving ions are known, then Faraday's law gives the mass of metal removed, m:

$$m = \frac{AIt}{zF} \tag{4.1}$$

where I is the current, t is the time of machining, and F is Faraday's constant. Thus, the rate of metal removal, $\dot{m}$, is

$$\dot{m} = \frac{AI}{zF} \tag{4.2}$$

It is often more convenient to work in terms of volumetric removal rates. If ρ_a is the density of the metal, from Equation (4.2) the volumetric removal rate, $\dot{v}$, is

$$\dot{v} = \frac{AI}{z\rho_a F} \tag{4.3}$$

Theoretical mass and volumetric removal rates for a range of elements are given in Table 4.1.

In Fig. 4.1, experimental removal rates for nickel are compared with theoretical values (based on divalent dissolution). For the

Table 4.1 Theoretical removal rates for a current of 1000 A

Metal	Atomic weight	Valency	Density (g/cm^3)	Removal rate (g/s)	Removal rate 10^{-6} m^3/s)
Aluminium	26·97	3	2·67	0·093	0·035
Beryllium	9·0	2	1·85	0·047	0·025
Chromium	51·99	2	7·19	0·269	0·038
		3		0·180	0·025
		6		0·090	0·013
Cobalt	58·93	2	8·85	0·306	0·035
		3		0·204	0·023
Niobium	92·91	3	9·57	0·321	0·034
(Columbium)		4		0·241	0·025
		5		0·193	0·020
Copper	63·57	1	8·96	0·660	0·074
		2		0·329	0·037
Iron	55·85	2	7·86	0·289	0·037
		3		0·193	0·025
Magnesium	24·31	2	1·74	0·126	0·072
Manganese	54·94	2	7·43	0·285	0·038
		4		0·142	0·019
		6		0·095	0·013
		7		0·081	0·011
Molybdenum	95·94	3	10·22	0·331	0·032
		4		0·248	0·024
		6		0·166	0·016
Nickel	58·71	2	8·90	0·304	0·034
		3		0·203	0·023
Silicon	28·09	4	2·33	0·073	0·031
Tin	118·69	2	7·30	0·615	0·084
		4		0·307	0·042
Titanium	47·9	3	4·51	0·165	0·037
		4		0·124	0·028
Tungsten	183·85	6	19·3	0·317	0·016
		8		0·238	0·012
Uranium	238·03	4	19·1	0·618	0·032
		6		0·412	0·022
Zinc	65·37	2	7·13	0·339	0·048

current densities shown (from about 60 to 186 A/cm^2) removal rate was independent of flow-rates in the range $0{\cdot}38 \times 10^{-3}$ to $1{\cdot}47 \times 10^{-3}$ m^3/s. Note that, as the current density is increased, the experimental rates become less than the theoretical rates. Explanations for this effect are discussed below in the section on current efficiency.

In the calculation of theoretical removal rates, confusion often arises if an incorrect valency state is attributed to the dissolving ion. For instance, nickel usually dissolves in the divalent state in nitrate and chloride solutions at low potential differences, but at higher potential differences it has been known to dissolve in the trivalent state [1].

The behaviour of copper serves as another example. This metal has been observed to dissolve in monovalent form in chloride solutions, and in the divalent state in nitrate solution [2].

Even this observation, however, needs qualification since the apparent valency of dissolution of copper changes with the mode of dissolution [3]. In nitrate and sulphate solutions copper dissolves in the active mode with an apparent valency of 2. For conditions of transpassive dissolution, the apparent valency has been found to lie between 1 and 2, e.g. 1·6 at 60 A/cm^2, the solution velocity being 6·86 m/s. This drop in apparent valency has been attributed to the simultaneous formation of monovalent copper reaction products. In chloride electrolytes, monovalent dissolution takes place in the active mode. For passivation conditions, the apparent valency increases, e.g. to a maximum value of 1·4 at 20 A/cm^2 and 6·06 m/s before reaching a limiting value of 1·2 at 40 A/cm^2 and 6·06 m/s. This increase is consistent with the copper dissolving partly in the divalent state. (Evidence of divalent dissolution was deduced from examination of the dissolution products. These appeared to be Cu_2O.)

4.2 Theoretical removal rates for alloys

Calculation of the removal rate for an alloy is more difficult than that for an element because the electrochemical equivalent of the alloy is not readily known. Although the electrochemical equivalents of the individual constituents of the alloy may be available, a difficulty arises in choosing a value for the electrochemical equivalent which is representative of the whole alloy. The problem has been considered by several workers [2, 4, 5]. The two methods which have been most commonly used are the 'percentage by weight' method and the 'superposition of charge' method.

Suppose first that an alloy consists of $X_A\%$ of element A, whose atomic weight is A_A and whose ions are known to dissolve in valency state z_A, of $X_B\%$ of element B with atomic weight and valency A_B and z_B respectively, and of $X_C\%$ of element C, etc.

4.2.1 *'Percentage by weight' method*

The sum of the chemical equivalents (A_i/z_i) of each element, i, in the alloy, multiplied by its respective proportion by weight X_i, gives a value for the chemical equivalent of the alloy, $(A/z)_{alloy}$. Thus

$$\left(\frac{A}{z}\right)_{alloy} = \frac{1}{100}\left[X_A\left(\frac{A_A}{z_A}\right) + X_B\left(\frac{A_B}{z_B}\right) + X_C\left(\frac{A_C}{z_C}\right) + \cdots\right] \quad (4.4)$$

The mass removal rate can then be found from Faraday's law:

$$\dot{m} = \left(\frac{A}{z}\right)_{alloy}\frac{I}{F}$$

4.2.2 *'Superposition of charge' method*

With this method, Faraday's law is used to calculate the amount of electrical charge required to dissolve the mass contribution of each constituent element to a defined mass (taken as 1 g) of the alloy.

Thus, for element A, the electrical charge required to dissolve the mass contribution $(X_A/100)$ g to 1 g of the alloy is

$$\frac{X_A}{100}\left(\frac{z_A}{A_A}\right)F \text{ coulomb}$$

Similarly, for element B, the equivalent electrical charge is

$$\frac{X_B}{100}\left(\frac{z_B}{A_B}\right)F \text{ coulomb} \quad (4.5)$$

and so on, for elements C, D, etc.

The electrical charge required to dissolve 1 g of the alloy is

$$\left(\frac{z}{A}\right)_{alloy} F \text{ coulomb} \quad (4.6)$$

Equating the sum of the charge for the elements [Equation (4.5)] to the charge for the alloy [Equation (4.6)], we obtain

$$\left(\frac{z}{A}\right)_{\text{alloy}} F = \frac{F}{100}\left[X_{\text{A}}\left(\frac{z_{\text{A}}}{A_{\text{A}}}\right) + X_{\text{B}}\left(\frac{z_{\text{B}}}{A_{\text{B}}}\right) + \cdots\right]$$

That is,

$$\left(\frac{A}{z}\right)_{\text{alloy}} = 100\bigg/\left[\frac{X_{\text{A}}}{(A_{\text{A}}/z_{\text{A}})} + \frac{X_{\text{B}}}{(A_{\text{B}}/z_{\text{B}})} + \cdots\right] \tag{4.7}$$

Example

Nimonic 75 has the constituent elements with the percentage by weight given in Table 4.2. The chemical equivalent for the alloy can

Table 4.2 Constituent elements (with their percentages by weight and chemical equivalents) of Nimonic 75

Element	Ni	Cr	Fe	Ti	Si	Mn	Cu
Percentage by weight (X)	72·5	19·5	5·0	0·4	1·0	1·0	0·5
Atomic weight (A)	58·71	51·99	55·85	47·9	28·09	54·94	63·57
Valency (z)	2	3	2	2	4	2	1
Chemical equivalent (A/z)	29·36	17·34	27·93	23·95	7·02	27·47	63·57
Product $X(A/z) \times 10^{-2}$	21·286	3·381	1·396	0·096	0·070	0·275	0·318
Quotient $X/(A/z)$	2·47	1·12	0·18	0·02	0·14	0·04	0·01

(Note: Nimonic 75 also contains 0·1% of carbon, which is inert)

now be calculated. By the first method, on the addition of the quantities

$$\sum_{\text{i}} X_{\text{i}}\left(\frac{A_{\text{i}}}{z_{\text{i}}}\right)$$

the chemical equivalent is calculated to be 26·8. By the second method, the addition of the quotients

$$100\Big/\sum_i \frac{X_i}{(A_i/z_i)}$$

yields a value of 25·1. Little difference is apparent in the values for each calculation. But for another alloy, Monel, whose constituents

Table 4.3 Constituent elements of Monel

Element	Ni	Cu	Fe	Mn	Si	C
Percentage by weight (X)	63	31·7	2·5	2	0·5	0·3
Chemical equivalent (A/z)	29·36	63·57	27·93	27·47	7·02	–
Product $X(A/z) \times 10^{-2}$	18·5	20·2	0·70	0·55	0·04	–
Quotient $X/(A/z)$	2·15	0·50	0·09	0·07	0·07	–

are given in Table 4.3, the calculations produce 39·9 and 34·7 respectively for its chemical equivalent. This discrepancy is indicative of the caution that is required in calculations by these methods. Moreover, their use in the prediction of machining rates for alloys requires a number of assumptions which again include knowledge of the valency states on dissolution of the constituent elements. The methods do not account for the effects of process variables, for

Table 4.4 Experimental and theoretical removal rates for Nimonic 75 alloy

	Machining rates (g min^{-1} cm^{-2})		
		theoretical	
Current density (A/cm^2)	observed	percentage by weight method	superposition of charge method
9·3	0·15	0·155	0·145
11·6	0·18	0·193	0·181
14	0·22	0·233	0·219
23·2	0·36	0·387	0·363

example, current density and flow-rate, on the rate of metal removal. As an example, however, some experimental results for Nimonic 75 are quoted in Table 4.4. The applied voltage was 12 V, and a 20% (w/w) NaCl solution, flowing at rates between $0{\cdot}38 \times 10^{-3}$ and $1{\cdot}47 \times 10^{-3}$ m^3/s, was used; no variation in removal rate due to rate of flow was observed over this range. Although there is a slight divergence, the table shows reasonable agreement between predicted and experimental removal rates.

4.3 Current efficiency

In the previous section, a difference in theoretical and experimental removal rates was noted. This result is often met in ECM. When the cause stems from the use of current for purposes other than metal removal, then a quantitative description can be obtained from *current efficiency* values. The current efficiency is defined as the ratio of the observed amount of metal dissolved to the theoretical amount predicted from Faraday's law, for the same specified conditions of electrochemical equivalent, current, etc.

If the metal ions dissolve in one valency form z, at current I, we can write

$$\text{Current efficiency} = \frac{\dot{m}}{(A/zF)I}$$

where $\dot{m}$ is now the observed mass removal rate corresponding to the specified experimental conditions.

It is often convenient to express the current efficiency in terms of a percentage ratio. For an efficiency of 100%, the entire current is used to dissolve the metal in accordance with Faraday's law. For zero efficiency, the current passes without metal dissolution.

Apparent current efficiency values, of course, can be affected by other factors. Since the valency states of ions can vary from solution to solution, or on alteration of a process variable, then clearly, the choice of the wrong valency can lead to an inaccurate estimate of the current efficiency. (From the example in Section 4.1, in which copper has been observed to dissolve in mono- and di-valent form, the estimated efficiency could be in error by 100%.) When alloys are machined, estimated values for their electrochemical equivalents must be used to estimate the current efficiency. We have seen that, with the usual methods of calculations, the valency values assumed for the individual constituents of the alloys can be a possible source of error.

Passivation can also reduce the current efficiency by reducing the metal removal rate. In other cases, grain boundary attack may occur causing removal of grains of metal from the anode surface by the traction forces of the electrolyte flow. Then, the metal removal rate and the corresponding current efficiency can be increased.

Gas evolution at the anode is another common reason for erroneous estimates of efficiency. In some cases, the generation of gas at the anode can be associated with an increase in current density. Thus, at low current densities, if little or no gas is evolved at the anode, the current efficiency should be high. But, at higher current densities, more gas will be evolved, more current will be used in its generation, and the current efficiency will be accordingly reduced. In other cases, however, an increase in current density leads to an increase in current efficiency. Moreover, for a specified set of process conditions, but with a change of electrolyte, widely different current efficiency values for one metal can be obtained. Examples to illustrate these points are set out below.

4.3.1 Effects of current density and solution flow on current efficiency

The results shown for nickel in Fig. 4.1 can now be presented in terms of current efficiency values (Fig. 4.2). Note that the current efficiency decreases slightly with an increase in current density. This trend has been reported for other metals in NaCl solutions.

Cuthbertson and Turner [6], who have machined Nimonic 80 in saturated NaCl, estimate, by the percentage by weight method, that the current efficiency decreases from 82% at 15·5 A/cm^2 to

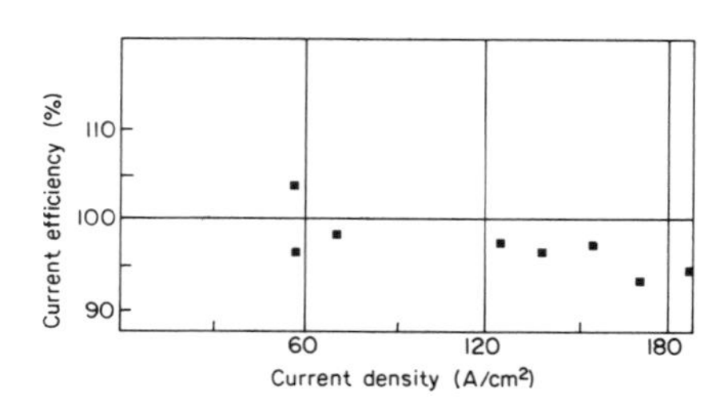

Fig. 4.2 Current efficiency as a function of current density for nickel

76% at 46·5 A/cm^2, the electrolyte velocity being 2·85 m/s. They also report a slight increase in current efficiency with electrolyte velocity from 74% at 1·26 m/s to 76% at 2·85 m/s (for a current density of 46·5 A/cm^2). The current efficiency for copper in 20%

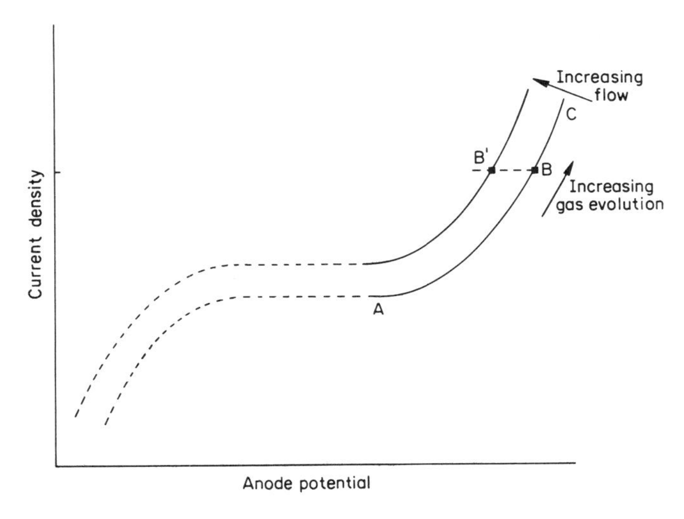

Fig. 4.3 Idealised polarisation curve

(w/w) NaCl also has been noted to increase with flow-rate up to a limiting value, e.g. at 6·96 A/cm^2, 92% for a flow-rate of 0·96 x 10^{-3} m^3/s, above which rate no further effect of flow-rate on efficiency was obtained [2]. During this investigation, no ECM was possible below a lower limiting flow-rate, owing to electric 'shorting' across the gap. (Severe mass transfer limitations on copper dissolution have been the subject of other more detailed studies [3].) A decrease in current efficiency with an increase in current density for mild steel in 2M and 4M NaCl has also been found by Mao [7], from about 99% at 26 A/cm^2 to 94% at 62 A/cm^2. He also reports that the cathodic efficiency for hydrogen generation is 100%.

These types of behaviour can be interpreted in terms of the polarisation curves which were introduced in Chapter 3. For convenience, an idealised curve is again given here (Fig. 4.3). In the region of A, only metal dissolution is supposed to occur. As the current density is increased along ABC, the anode potential also increases, and the higher energy then made available allows other, higher-energy electrode reactions, e.g. oxygen evolution, to occur.

Since the gas evolves in preference to metal dissolution, and in progressively greater quantities along ABC, the current efficiency for metal removal decreases. For a given current density, say at B, the effect of an increase in flow-rate, however, is the translation of the point B to B′. The latter point corresponds to a lower anode potential, so conditions are again less conducive to gas evolution. Accordingly, the current efficiency for metal removal is increased. This behaviour is not necessarily characteristic of all metal–electrolyte combinations. For instance, for a nickel–chromium alloy machined in $NaClO_3$, $NaNO_3$, $NaNO_2$, Na_2SO_4, and $Na_2Cr_2O_7$ electrolytes, an increase in current efficiency with current density has been reported [8].

Similar observations have been made by Bergsma [9] for steel in $NaClO_3$, and by Mao for mild steel in $NaClO_3$, $NaClO_4$, and $NaNO_3$ solutions [7, 10]. The latter's experiments are discussed below.

4.3.2 *Effects of different electrolytes on current efficiency*

(*a*) *Mild steel and $NaClO_3$ solution*

Current efficiency studies for both the dissolution of metal and the generation of gas in $NaClO_3$ electrolyte are summarised in Tables 4.5

Table 4.5 Current efficiency values for mild steel machined in $NaClO_3$; air atmosphere; electrolyte flow-rate $0{\cdot}03 \times 10^{-3}$ m^3/s (after Mao [10])

$NaClO_3$ concentration (M)	Cathode material	Apparent current density (A/cm^2)	Current efficiency for H_2 generation (%)	Divalent iron dissolution (assumed) Current efficiency (%)	pH change from	pH change to
4·5	Brass	47	98·2	74·8	7·7	7·4
4·5	Pt	47	98·1	80·7	—	—
4·5	Brass	47	96·7	82·9	—	—
4·5	Pt	47	99·3	78·6	8·1	6·1
4·5	Brass	26	99·0	58·6	7·9	6·5
4·5	Pt	26	99·0	60·7	7·3	6·3
4·5	Brass	26	99·8	62·7	8·5	6·3
4·5	Pt	26	98·6	68·8	8·1	5·7
2·0	Brass	47	98·7	60·4	7·9	6·4
2·0	Brass	47	99·9	67·1	7·5	5·2
2·0	Brass	37	98·3	46·3	7·0	6·0
2·0	Brass	37	100·9	59·7	7·4	6·8

and 4.6. These experiments were carried out in cell atmospheres of both air (Table 4.5) and nitrogen (Table 4.6). A nitrogen atmosphere was also used so that the quantity of anodic oxygen would be detected, as well as metal removal.

Table 4.6 Current efficiency values for mild steel machined in $NaClO_3$; nitrogen atmosphere; brass cathode; electrolyte flow-rate 0·03 x 10^{-3} m^3/s (after Mao [10])

				Divalent iron dissolution (assumed)			
$NaClO_3$ concentration (M)	Apparent current density (A/cm^2)	Current efficiency for H_2 generation (%)	Current efficiency for O_2 generation (%)	Current efficiency (%)	Total current efficiency for O_2 and Fe (%)	pH change from	to
4·5	47	98·6	24·8	78·6	103·4	8·4	5·7
4·5	37	97·2	27·4	72·6	100·0	8·2	6·7
4·5	26	98·8	35·3	64·3	99·6	7·8	6·8
4·5	12	100·1	57·9	41·7	99·6	8·2	7·2
2·0	47	98·7	37·5	63·4	100·9	6·2	5·8
2·0	37	98·1	39·5	61·0	100·5	6·4	6·2
2·0	12	101·2	80·8	19·1	99·9	6·5	6·2

Several significant conclusions can be made from these experiments. First, the results show that the cathodic reaction is essentially hydrogen evolution and that the current efficiency is not influenced by the two cathode materials. Next, from Table 4.6, the formation of oxygen takes place at the anode in addition to metal dissolution (probably ferrous iron). Also, during these experiments, NaCl was formed or $NaClO_3$ reduced, the probable overall chemical reaction being $6Fe^{++} + NaClO_3 + 3H_2O \rightarrow 6Fe^{+++} + NaCl + 6OH^-$.

Mao claims that these results are largely determined by an iron oxide film on the anode surface [10]. The film is assumed to be porous to allow the rapid passage of cations through it, with the pores small and randomly distributed. Nevertheless, the film also acts as a barrier which hinders the movement of the reaction product (the ferrous ion). The rate of metal removal is thereby decreased, the extent of decrease being dependent on the film thickness. With diminished metal removal, part of the current is used for the generation of oxygen, the possible electrode reaction being

$$2H_2O \rightarrow O_2 + 4H^+ + 4e$$

The properties of the anode film can be explored further with the aid of Table 4.6 and Fig. 4.4. Their data show that the current efficiency for metal removal increases with current density, with a corresponding decrease in current efficiency for oxygen generation.

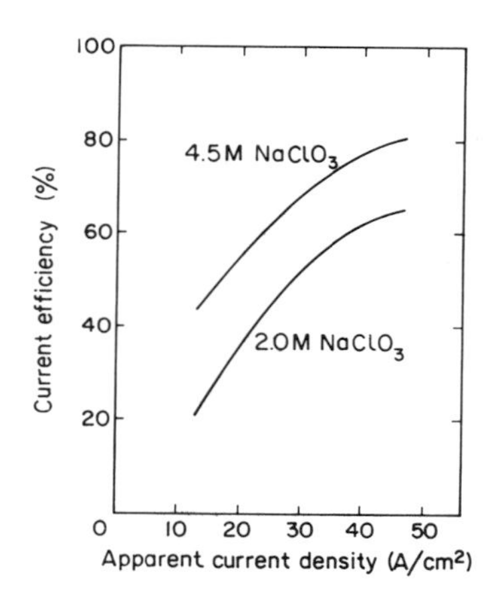

Fig. 4.4 Current efficiency for divalent iron dissolution as a function of current density (after Mao [10])

Mao suggests that as the current density is increased the thickness of the oxide film becomes thinner, so that the metal removal rate is increased. Since iron oxide is known to be more soluble at low pH, the decrease in film thickness could be caused by a locally lower pH value in the vicinity of the anode surface. That condition could be produced by the formation there of oxygen and ferrous, or ferric, hydroxide.

One other result is of interest from that study. Figure 4.5 shows that the addition of $NaClO_4$ to the main $NaClO_3$ electrolyte increases the current efficiency. The oxide film clearly must have been influenced by the effective concentration of the new electrolyte solution. The main effect of the additive is to lower the oxidising power (or the useful concentration) of the $NaClO_3$ solution. At high concentrations of the additive, however (above 0·25M), another feature of $NaClO_4$ becomes evident. At these concentrations, the $NaClO_4$ itself attacks the oxide film, causing pitting. At concentrations of about 0·5M, severe grain boundary attack leads to current efficiencies in excess of 100%. These results will be discussed further in relation to surface finish.

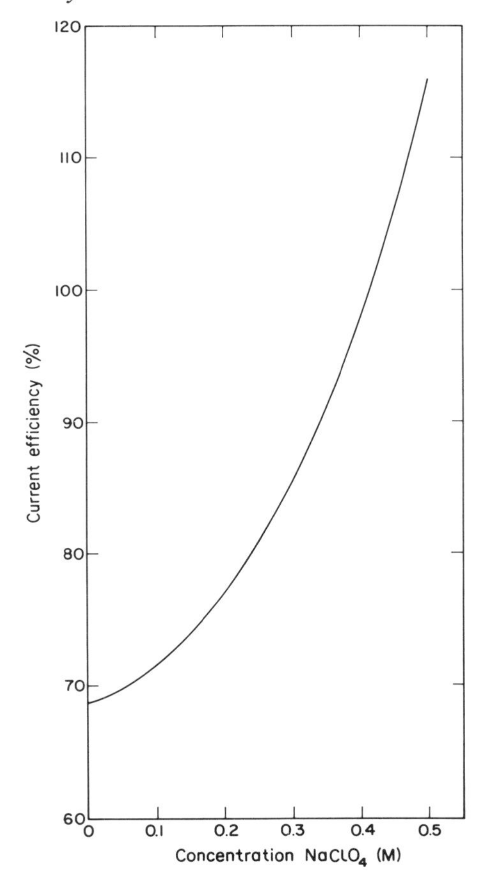

Fig. 4.5 Effect on current efficiency of ClO_4 ions added to 3M $NaClO_3$ electrolyte; current density 37 A/cm^2 (after Mao [10])

(*b*) *Mild steel and $NaClO_4$ and $NaNO_3$ solutions*

The above studies have led Mao to investigate the ECM of mild steel in other electrolytes with different oxidising strengths [7]. For example, he has found that, in 2·0M and 4·5M $NaClO_4$ solution and for current densities ranging from 26 to 47 A/cm^2, the cathode current efficiency for hydrogen generation is approximately 100%. At the anode, the current efficiencies for metal removal and oxygen generation are about 90% and 4 to 8% respectively, the total anode current efficiency being about 96%. The smaller amount of oxygen generated in $NaClO_4$ compared with $NaClO_3$ is probably due to a weaker oxide film in the former solution and to its removal by the powerful perchlorate ions. Nevertheless, since a little oxygen is still formed during ECM, some residual oxide film must still be present. Total anode current efficiencies lower than 100% suggest that some oxygen is also used to oxidise the ferrous product of reaction. (Additional evidence has suggested that the final product is Fe_3O_4.)

Table 4.7 Current efficiency values for mild steel machined in $NaNO_3$ and nitrogen atmosphere (after Mao [7])

$NaNO_3$ concentration (M)	Cathode	Apparent current density (A/cm^2)	Current efficiency for H_2 generation (%)	Divalent iron removal (assumed): Current efficiency for O_2 generation (%)	Divalent iron removal (assumed): Current efficiency (%)	Total current efficiency for O_2 and Fe (%)	pH change: from	pH change: to
4·5	Pt	47	1·1	54·7	33·2	87·9	5·3	11·4
4·5	Brass	47	0	55·9	33·0	88·9	6·3	11·3
4·5	Brass	26	0	50·6	36·8	87·4	6·2	11·2
4·5	Brass	12	0	96·4	1·9	98·3	5·7	11·3
2·0	Pt	47	11·6	83·7	11·7	95·4	6·0	11·4
2·0	Brass	47	0	84·5	11·5	96·0	6·1	11·4
2·0	Brass	26	0	96·9	2·7	99·6	6·1	11·5

The performance of $NaNO_3$ electrolyte is quite different from that of the previous electrolytes (Table 4.7). No hydrogen is generated at a brass cathode, and only a small amount of the gas is formed when the cathode material is platinum. The alternative cathodic reaction appears to be reduction of nitrate, in which case ammonia, hydroxylamine, and nitrite are detected in the $NaNO_3$ solution after ECM.

The current efficiency for metal removal is much lower than that for $NaClO_3$, and most of the current is used for the evolution of oxygen at the anode. From these results, the oxide film in $NaNO_3$ would appear to be less porous than that in $NaClO_3$. These experiments also show that the current efficiency for iron removal in $NaNO_3$ increases with electrolyte concentration and with current density. Explanations for these effects are given above in the section on $NaClO_3$. Since the total anode current efficiency for iron removal and oxygen generation is less than 100%, either some iron may be dissolved as ferric ions, or some oxygen is used in oxidising the products of reaction to produce ferric hydroxide.

(*c*) *Mild steel and $NaNO_3$ solution containing additives*

Chikamori and Ito [11] have also reported that the current efficiency for mild steel machined in $NaNO_3$ solution increases with increasing current density, and that the metal removal rates are significantly less than those achieved in NaCl and $NaClO_3$ electrolytes. They believe that additives to the main solution which have oxygen-containing anions should give improved performance at higher current densities. Some results, presented in Table 4.8, do show a general

Table 4.8 Removal rates for mild steel; electrolyte velocity 8·3 m/s; temperature 40°C (after Chikamori and Ito [11])

Electrolyte		Metal removal rate (mg/C): Current density 25 A/cm²	Current density 7·5 A/cm²	Current density 2·5 A/cm²	Specific conductivity (ohm⁻¹ cm⁻¹ at 20°C)	pH
100 g/l NaCl		0·286	0·297	0·292	0·119	4·7
300 g/l $NaClO_3$		0·257	0·200	0·098	0·132	6·2
200 g/l $NaNO_3$		0·104	0·011	0·004	0·126	7·5
300 g/l $NaNO_3$		0·142	0·022	0·006	0·156	8·2
Additive (g/l) to 200 g/l $NaNO_3$						
$(NH_4)_2SO_4$	30	0·181	0·064	0·046	0·143	5·6
	100	0·197	0·084	0·048	0·180	5·4
$NaBrO_3$	30	0·176	0·027	0·004	0·127	8·3
	100	0·200	0·036	0·008	0·135	8·6
$KBrO_3$	30	0·170	0·028	0·006	0·131	8·1
	100	0·200	0·053	0·006	0·146	7·4
Na_2SO_4	30	0·163	0·053	0·030	0·133	8·2
	100	0·179	0·073	0·036	0·142	7·7
$Na_2Cr_2O_7, 2H_2O$	30	0·162	0·027	0·008	0·132	3·7
	100	0·184	0·040	0·010	0·143	3·9
$NaClO_3$	30	0·160	0·042	0·008	0·133	7·3
	100	0·272	0·182	0·030	0·148	7·9
NH_4NO_3	30	0·150	0·013	0·008	0·145	6·2
	100	0·176	0·025	0·022	0·192	6·0
$(NH_4)_2S_2O_8$	1	0·140	0·037	0·041	0·126	2·6
NaCl	30	0·091	0·027	0·107	0·147	7·6
	100	0·033	0·032	0·296	0·184	8·0

increase in removal rates. However, the addition of NaCl has the effect of increasing removal rates at low current densities and decreasing the rates at higher current densities. In this case, an undesirable feature is a poorer surface finish.

4.4 Power efficiency

Although current efficiency is a commonly used quantity in ECM work, a more accurate evaluation of the process should include an estimate of its electrical power efficiency.

The estimate can be based on the calculation of the power required to pass a specified current across the machining gap. Since the current is fixed, only the voltage component of the power needs consideration. Now, in Chapter 3, the voltage drop across an ECM cell was shown to be composed of a main resistance drop (*IR*) across the gap, with other supplementary contributions which include the overpotentials at both electrodes. The resistance drop being inversely proportional to the electrolyte conductivity, a study of some relevant features of the latter quantity should show means of lessening power requirements in ECM.

One characteristic which is often utilised in practice is the rapid increase in electrolyte conductivity with temperature (40 to 60°C are common working temperatures). For some electrolytes, the conductivity can be increased two-fold for a relatively small change in temperature (e.g. from 20°C to 40°C). Therefore, for the same operating conditions, the voltage, and hence the power required, are decreased by equivalent amounts. In these cases, substantial increases in the power efficiency of the process are achieved.

Conductivity values, of course, also vary greatly with concentration, from one electrolyte to another, and even on the addition to one main electrolyte of a small quantity of some other solution. For instance, suppose that 25 V is required for the passage of 1 kA in 10% NaCl electrolyte (conductivity, $\kappa_e = 0{\cdot}1$ ohm^{-1} cm^{-1}) at 18°C. In 20% NaCl electrolyte ($\kappa_e = 0{\cdot}2$ ohm^{-1} cm^{-1}), 12·5 V are needed, whilst in a solution of 15% NaCl containing 1% Na_2CO_3, 15 V are required. In the first case, the power required is 25 kW, in the second, 12·5 kW, and in the third case 15 kW.

It should be remembered, however, that electrolytes of extremely high conductivity tend to be very acidic or alkaline. Their use in practice is limited; for instance, with the acidic type, corrosion problems make them difficult to handle. For this reason, neutral solutions, which have lower conductivities, are more common.

Also, the presence of the products of machining within the gap can influence the effective conductivity of any solution. Their detrimental effects can usually be lessened by an increase in the electrolyte flow-rate.

4.5 Types of surface finish

Many of the factors which influence the rate of dissolution also affect the manner in which metal is removed from the anode, and hence they partly determine the surface finish. Of these factors, the anode potential and current density play a major part. Their rôle can be usefully studied from polarisation curves. Information which these curves supply on the different grades of surface finish likely to be found in ECM, and examples where they are met in

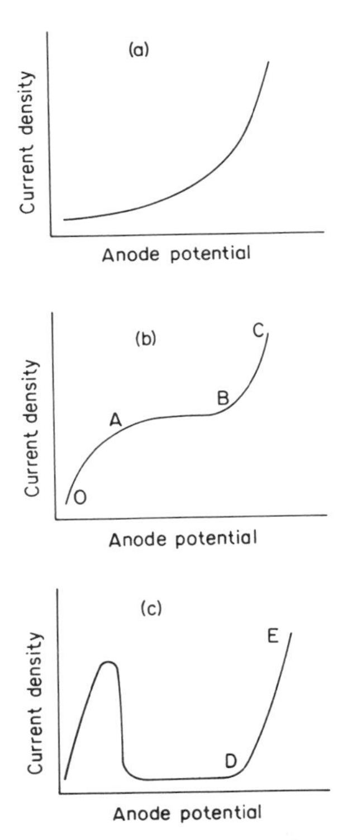

Fig. 4.6 Anode potential–current density polarisation curves

practice, are now discussed. During this discussion, it will become clear that anodic films greatly influence the quality of surface finish.

4.5.1 Etching

When etching occurs, the anode potential–current density curve usually has the form shown in Fig. 4.6(a). An explanation for etching, relevant to ECM, has been given by Boden and Brook [12]. Suppose that the individual atoms of a metal can be assumed to be spherical and simply arranged, and that the individual crystals of the metal, which are formed by the atomic lattice, can take almost any orientation of that lattice [Fig. 4.7(a)]. Slight, local variations in dissolution rates then occur, with metal dissolving more rapidly from areas which have a wider atomic spacing. Eventually, an uneven surface is produced, the exposed face having the structure of the most closely spaced atoms. On further dissolution, these close-packed layers of atoms are also removed, and the grains become disjunctive. If the grain boundaries are dissolved at a faster rate than their surrounds, as indeed often happens, entire grains can be dislodged from the metal surface. The etched finish of the metal arises from the non-specular reflection of light from the crystal faces so dissolved

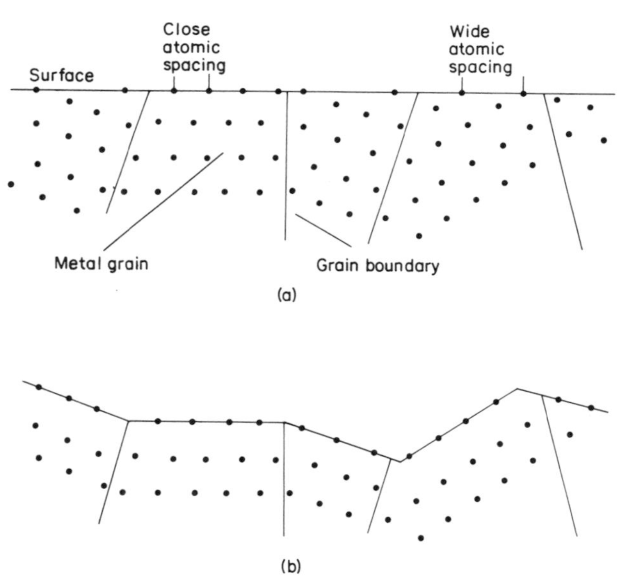

Fig. 4.7 Structure of a metal surface produced by (a) mechanical polishing, (b) mechanical polishing, then etching

at different rates. The anodic processes which lead to etching are commonly said to be 'activation-controlled'. We have already noted that Tafel's equation is usually assumed to be valid for etching conditions.

A detailed study of the active dissolution of copper in 2M KNO_3 has emphasised that crystal orientation has considerable bearing on the development of etched surface finishes [13]. Dissolution here proceeded by the removal of atoms from close-packed planes of the crystal lattice. This manner of metal removal led to a structure of surface ridges, formed by sub-microscopic steps on the metal surfaces. From this configuration, the etched appearance of the metal was obtained. This investigation also confirmed that flow-rate has no effect on this type of finish.

Etched matt finishes of approximate roughness 0·9 to 1·4 μm at current densities and electrolyte flow-rates of 15·5 A/cm^2 and 28 m/s respectively have been produced on Nimonic 80 machined in saturated NaCl solution [6].

4.5.2 Polishing

Figure 4.6(b) shows the usual shape of the polarisation curve for polishing conditions. Over the region OA, etching still occurs. But over the region AB, the anode surface becomes covered with a film and a polished finish results.

Ni^{++}	O^{--}	Ni^{++}	O^{--}	Ni^{++}	O^{--}
O^{--}	Ni^{++}	O^{--}	Ni^{++}	O^{--}	Ni^{++}
Ni^{++}	O^{--}	Ni^{++}	O^{--}	Ni^{++}	O^{--}
O^{--}	Ni^{++}	O^{--}	Ni^{++}	O^{--}	Ni^{++}

(a)

Ni^{++}	O^{--}	Ni^{++}	O^{--}	Ni^{++}	O^{--}
O^{--}	Ni^{++}	O^{--}	Ni^{+++}	O^{--}	Ni^{++}
Ni^{+++}	O^{--}		O^{--}	Ni^{++}	O^{--}
O^{--}	Ni^{++}	O^{--}	Ni^{++}	O^{--}	Ni^{++}

(b)

Fig. 4.8 (a) Ideal structure of NiO. (b) Real structure of NiO showing vacancy and trivalent nickel

The nature of the mechanism controlling electropolishing is still not clear, although the random removal of atoms from the surface seems the main cause. That process can be explained in terms of a 'solid defect structure' on the anode surface [12]. (A defect structure is an arrangement of ions which includes holes or vacancies in the ionic lattice by means of which ions migrate.) The structure of nickel oxide provides a convenient example. Its simple, complete structure, shown in Fig. 4.8(a) gives the formula NiO. However, from chemical analysis, its true formula can be shown to be approximately $Ni_{1-r}O$ where $r \simeq 0{\cdot}05$. That is, there is a minor deficiency of metal. The discrepancy can be explained by the formation of a few Ni^{+++} (nickelic) ions and an equivalent number of vacancies, so that electrical neutrality is maintained. The structure of the oxide in this condition is given in Fig. 4.8(b).

Suppose now that, on the dissolution of a nickel anode, the metal becomes covered with such a layer of nickel oxide. Dissolution will proceed by means of the transport of ions across the oxide. We assume further that dissolution conditions (for example, the formation of a nickelic ion) also lead to the presence of a vacancy in the lattice at the oxide–solution interface. Movement of this vacancy towards the metal–oxide interface will be equivalent to the movement of metal ions in the opposite direction. On the arrival of the vacancy at the metal–oxide interface, an atom from the metal enters the vacancy and becomes charged. This arrival of the vacancy at the interface is a random process and is therefore independent of crystal orientation. The metal, accordingly, also dissolves in a random fashion, and a level surface is obtained. An electropolished surface finish will be achieved, provided that the film's properties facilitate sufficiently the movement of metal ions through the lattice. The electron conductivity of the film must also be sufficiently high to permit the process to proceed at low anode potentials.

Although these properties are not normally encountered in oxide films, some films which do contain anions other than oxygen do have the requisite properties. Boden and Brook consider that these foreign anions penetrate the film under the influence of the electrostatic field across the oxide–solution interface, thus distorting the film. They also believe that the film must be assumed to be solid, to explain the random removal of metal, and that the assumption of a lattice distorted by foreign anions in the solution is necessary for the explanation of rapid ionic transport through the film.

Beyond the solid film, a viscous layer is maintained due to the high concentration of anions and cations. The thickness of this layer, of course, depends on the rate of diffusion into the bulk of the electrolyte and on the electrolyte flow-rate (see Chapter 3). The formation of the layer is also influenced by the current density. At low current densities, the cations may diffuse away from the anode surface at a rate greater than that at which they are produced by dissolution. The concentrated layer, essential for film formation, is not produced, and etching results [Fig. 4.6(a)]. At high current densities, the process is diffusion-controlled, and a current density plateau region is obtained, as indicated in Fig. 4.6(b). The anode surface is then electropolished. Cuthbertson and Turner [6] have reported that Nimonic 80 polishes at 46·5 A/cm^2 and 28·5 m/s in saturated NaCl (surface roughness 0·2 μm CLA). They conclude that ECM must have been controlled by a viscous diffusion layer on the anode surface.

4.5.3 Polishing with pitting

At higher potentials [along BC in Fig. 4.6(b)] the polishing action is accompanied by pitting of the surface. The latter effect can be attributed to the onset of gas evolution at the anode surface, and to rupture (in some way not yet fully established) of the anodic layer by the gas. Small, fixed active sites form on the metal surface which give rise to pitting.

4.5.4 Passivation

When passivation occurs, the anodic polarisation curves take the characteristic shape, shown in Fig. 4.6(c).

Hoar [14] considers the principal action in passivation to be the very rapid formation of an oxide layer which becomes firmly attached to the metal and which forms a barrier between it and the solution. The solid film so produced has a low ionic conductivity so that the rate of dissolution of metal is decreased (partial passivation) or made negligible (total passivation). It is thought that passivating films can be either very poor electron conductors [14] or good electron conductors [15].

Total passivation with anodic gas evolution, encountered during polarisation studies of carbon steel in $NaNO_3$ electrolyte, has been attributed to an electronically conducting passive layer on the anode surface [16]. But an electronically non-conducting passive film is the proposed reason for total passivation without gas generation in further work with this metal in aqueous potassium fluoride electrolyte.

The passive oxide film, which makes titanium so useful as a corrosion resistant material, renders the ECM of this metal very difficult. With simple chloride and nitrate electrolytes, high applied potential differences, e.g. 50 V, are often required to achieve machining although the passive films are then broken only at weak points causing deep attacks at grain boundaries. Even so, the power efficiency of the process is greatly reduced, and high electrolyte flow-rates are required to maintain reasonable temperatures within the gap.

It has been claimed that the voltage required to break down the passive film can be reduced by the use of bromide and iodide electrolytes, e.g. 3M NaBr and NaI [1, 17]. Even then, however, a severely pitted surface finish is obtained. An equally poor finish has been found with titanium alloy dissolved in 3M NaCl [17]. The achievement of machining in that case has been attributed to possible hydrogen gas generation at the anode. (Evidence for such evolution stemmed from high acidity levels of the solution in the neighbourhood of the anode in comparison with the neutral pH values of the bulk solution.) The inferred hydrogen ion activity and a consequent reaction with the solid anodic film of TiO_2 is thought to have eventually caused a breakdown of this film. ECM is proposed to take place then by localised attack at active sites on the anode.

Tungsten carbide is another example in which passivation hinders machining. The cobalt in this metal forms a passive oxide film which has to be broken if machining is to be achieved. Although sodium nitrate is known to dissolve the basic tungsten carbide metal, the addition of an aggressive anion is helpful in breaking down the passive film formed by the cobalt.

Partial passivation, characterised by a thick, tenacious black film on the anode surface, is thought to be the cause of the coarse finish (surface roughness in excess of 5·2 μm CLA) and low current efficiency (about 40%) obtained with cast iron machined in 20% (w/w) NaCl at current densities from about 7·7 to 15·5 A/cm^2 [18], flow-rates in the range 0·4 to 1·4 x 10^{-3} m^3/s being used.

Normalised, and quenched and tempered steels, with carbon contents of 0·99% and 1·26%, machined in 20% (w/w) NaCl at 25 A/cm^2 and 15 V, have also yielded low current efficiencies and dull granular finishes [16]. The performance of these metals appears to be influenced by a slightly soluble resistance layer which favours partial passivation. (The values of the other process variables in these experiments were: gap width, 0·75 mm; inlet temperature, 18 to 27°C; electrolyte velocity, 16·5, 27, and 45 m/s).

4.5.5 *Transpassivation*

On further increase in potential, conditions at an anode may change from the passive to the transpassive state. In the polarisation curves, this change is seen as a marked increase in current density [along line DE in Fig. 4.6(c)]. It is caused by the anodic oxidation of the original, slightly soluble passive oxide film to a soluble form. With the occurrence of transpassivity efficient ECM takes place. A good example is provided from the work by LaBoda and McMillan [19] on the ECM of low-alloy nickel–chromium steel in 450 g/l $NaClO_3$ solution. Using current densities of 39 to 116 A/cm^2 and a (maximum) flow-rate of $0{\cdot}63 \times 10^{-3}$ m^3/s, they achieved surface finishes of 0·1 μm. Later work established that these experiments must have been conducted in the transpassive region.

4.5.6 *Transition from active to transpassive dissolution*

Transpassive machining conditions can also be reached from an initial, active state of dissolution. Then, the transition becomes evident by the change from a dull, etched finish to a smooth, bright

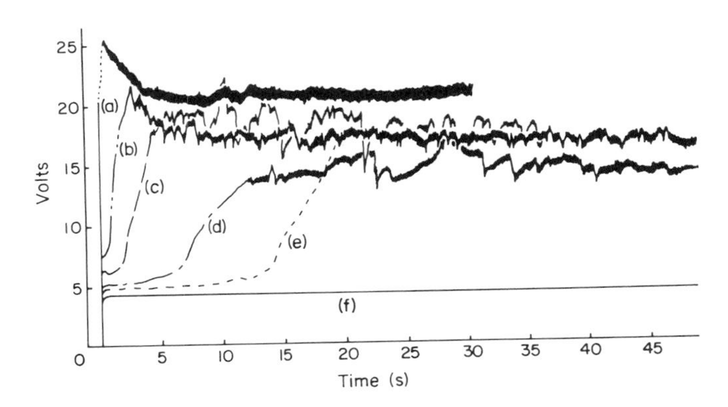

Fig. 4.9 Cell voltage transients measured in 2N KNO_3 with flow-rate 0·3 m/s; current density (A/cm^2): (a) 16·5, (b) 8·3, (c) 5·6, (d) 4·4, (e) 4·0, (f) 3·4 (after Kinoshita et al. [3])

one. Transient cell voltage–current measurements can also reveal this transition. Figure 4.9 shows such behaviour for galvanostatic dissolution of copper in 2N KNO_3 [3].

At the current density of 3·4 A/cm^2, the metal dissolves indefinitely in the active mode. At a higher current density and after a transition time with active dissolution, the dissolution process changes to the transpassive mode. The cell voltage is then about 10 to 20 V higher than that for the active mode. The critical current density at which this change occurs is increased by increased flow-rate, the transition period being shortened by increased current density and decreased flow-rate. Figure 4.9 also shows that voltage fluctuations occur in the transpassive mode.

Since the transition from active to transpassive dissolution is influenced by flow-rate and current density, a short digression on the possible cause of this effect is of interest [13].

First, the expression is introduced which describes the experimental mass transfer rates at small distances from the leading edge of the electrode and for fully developed velocity profiles:

$$\mathrm{Nu} = 0{\cdot}28\ \mathrm{Re}^{0{\cdot}58}\ \mathrm{Sc}^{1/3}\left(\frac{d_{\mathrm{h}}}{L}\right)^{1/3} \tag{4.8}$$

[cf. Equation (3.33)]. Here, Nu is defined by

$$\mathrm{Nu} = \frac{J_{\mathrm{l}} d_{\mathrm{h}}}{zFC_{\mathrm{b}}D} \tag{4.9}$$

in the usual notation.

Next we note that there is some indication from experiments that, at high dissolution rates, the product of the metal ion concentration and anion concentration might exceed the solubility product [cf. Chapter 3, Section 3.1(f)] [13]. The onset of this condition could lead to the precipitation of salt crystals, thereby initiating passivation.

For known hydrodynamic conditions, the Nusselt number corresponding to this condition may be calculated from Equation (4.9). Then, from Equation (4.8) and from the experimentally observed passivation current densities, J_{p}, an estimate can be made of the interfacial concentration of ionic dissolution product, C_{s}:

$$C_{\mathrm{s}} = \frac{J_{\mathrm{p}} d_{\mathrm{h}}}{zFD\mathrm{Nu}_{\mathrm{p}}} \tag{4.10}$$

Table 4.9 Calculations of interfacial concentration of copper salt at onset of transpassive dissolution (after Kinoshita et al. [3]) (Values used in calculations: $D = 5 \times 10^{-6}$ cm²/s; $\nu = 1$ mm²/s; d_h (= $2h$) = 1 mm; L = 0·53 mm; z = 2; $F = 9{\cdot}65 \times 10^4$ As)

Electrolyte	Electrolyte velocity (m/s)	Reynolds number	Passivation current density (A/cm²)	Interfacial concentration (mole/l)
2N KNO_3	0·30	510	3·8	4·5
	2·00	3 390	8·5	4·6
	6·27	10 600	31·0	8·5
1N K_2SO_4	0·50	730	1·5	1·4
	2·00	2 920	4·1	2·2
	6·86	10 000	8·5	2·2
1N H_2SO_4	0·50	770	3·6	3·4
	2·00	3 090	4·7	2·5
	6·86	10 600	11·8	3·0
2N KCl	0·50	820	0·4	0·8
	2·00	3 270	1·2	1·2
	6·06	10 000	2·5	1·2

where Nu_p is the Nusselt number associated with J_p. Calculated concentrations for a range of electrolytes, flow-rates, and passivation current densities are given in Table 4.9. In addition, solubilities of copper salts are presented in Table 4.10. Qualitative comparison of the two tables suggests that the onset of transpassive dissolution is approximately coincident with the limiting transport of dissolved reaction products by convective diffusion. Although similar behaviour has been known to occur in nitrate and sulphate solutions, in KCl no sharp transition from active to transpassive conditions

Table 4.10 Solubilities of copper salts (room temperature) (after Kinoshita et al. [3])

Salt	Solvent	Solubility (mole/l)
Cupric nitrate	H_2O	7·0
Cupric sulphate	H_2O	1·4
	1N H_2SO_4	1·1
Cuprous chloride	2N KCl	1·0
	3N KCl	1·9

has been found. Instead, periodic voltage oscillations in the transition region have been observed. The anodic behaviour in this chloride electrolyte is thought to be much more complicated, since various complexes may be formed depending on the local chloride concentration. For example, owing to the effect of chloride concentration on the solubility of cuprous ion, the formation of precipitate layers is dependent on the rate of arrival of anions at the interface as well as on the rate at which cations depart from the interface. Consequently, the mass transport of chloride ions may also be a limiting factor. Nevertheless, calculations have shown that for copper dissolution in 3N KCl the rate-limiting concentration difference for the onset of transpassivation (3·6 mole/litre) is of the same order as the solubility of CuCl in that solution (1·9 mole/litre). It is possible that, for some conditions, passivation in chloride solutions is caused by a process of precipitation which is similar to that for sulphate and nitrate solutions.

4.5.7 Periodic phenomena

The abrupt rise in anode potential which occurs for transpassive dissolution conditions appears to be linked with a thin compact layer on the anode surface.

Further investigations of the nature of anodic layers have indicated that, while their formation is initiated by mass transport limited processes, their removal also involves other factors. For example, the removal of these layers is known to become accelerated at higher current densities and flow-rates, and the rate at which they are removed can exceed their rate of formation at lower current densities. The current density effect may have one or several causes. First, with an increase in current density, greater power dissipation probably takes place through this layer, thereby decreasing its effective resistance. The consequent resistance heating may produce stresses within the layer which lead to its periodic rupture. Alternatively, the influence of current density may be caused by higher local temperatures which increase the rates of chemical dissolution, or by the collection of larger concentrations of hydrogen ions at the surface. The effect of flow-rate may be simply related to increased mechanical removal of solid products, or to increased rate of mass transport of dissolved products.

Periodic phenomena associated with unstable surface films occur for many metal–electrolyte combinations. In potentiostat work, these phenomena are seen as current density oscillations; in galvanostatic studies, they take the form of electrode potential fluctuations. The films may be salt films formed before the onset of passivation, or the films arising from the electropolishing process, or the films associated with transpassive dissolution.

For instance, Postlethwaite and Kell [20] have observed periodic potential fluctuations during galvanostatic anodic dissolution of iron in 20% (w/v) NaCl solution. In their experiments, Tafel behaviour was observed up to a critical current density (200 mA/cm^2) above which resistance to dissolution increased. This effect was seen as an abrupt increase in slope of the anode potential–current density curve. In this region the oscillations in potential took place. They were attributed to the presence of an unstable, porous, non-protective film, which was found to contain 70% iron.

Similar behaviour has been encountered during potentiostat work on carbon steels in 20% NaCl solution [16]. Again the oscillations appear to have been caused by the observed periodic rupture of a black film on the anode surface.

The nature of such oscillations has been studied during the galvanostatic dissolution of copper in $NaClO_3$ electrolyte [21]. Figure 4.10 shows the different types of potential oscillations which arise with increasing current density. Note that the period of oscillation decreases, and the shape of the cycle changes, with increasing current density. The period also has been found to increase slightly with increasing flow-rate.

Low current density studies (below 5 A/cm^2) have shown that any single period of oscillation has two characteristic phases, I and II, say. The length of phase I decreases as the current density is increased,

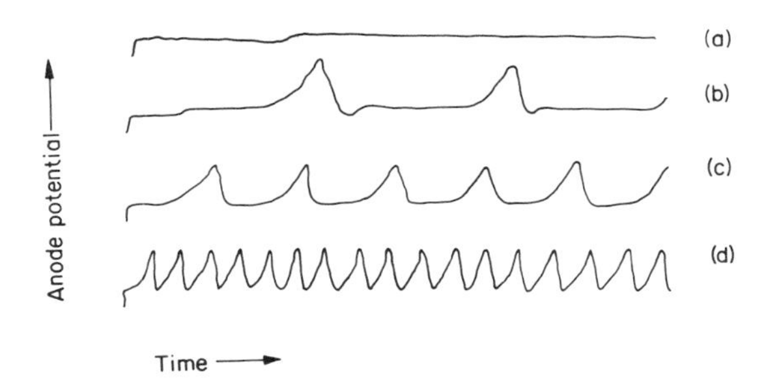

Fig. 4.10 Schematic representation of variations in form of potential oscillation: (a) J (A/cm^2) $< 0{\cdot}03$; (b) $0{\cdot}5 < J < 5$; (c) $5 < J < 10$; (d) $J > 10$ (after Cooper et al. [21])

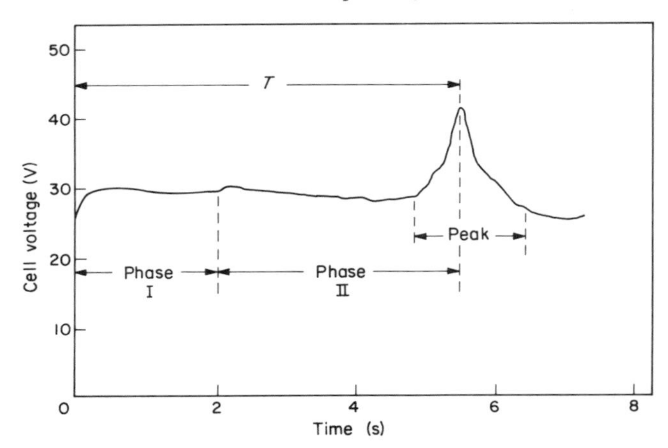

Fig. 4.11 Characteristic phases of a single period of oscillation (after Cooper et al. [21])

and becomes negligible at current densities above 5 A/cm^2. The characteristics of the second phase are a region of low potential and a potential peak. These phases are illustrated in Fig. 4.11 for copper dissolved in unstirred 2N $NaClO_3$ solution at 1·33 A/cm^2. This investigation showed that the potential cycle reflects the periodic formation and rupture of surface films. In phase I, a strongly adherent surface film of cuprous oxide was formed. In phase II the film was weakly adherent and it appeared to contain cupric chlorate and its basic salts as well as cuprous oxide. As shown in Fig. 4.12, the film thickness increases with time throughout the low and peak potential regions. Its rupture and explosion from the surface of the film substance are coincident with the sudden drop from the peak potentials.

It has been suggested that, in neutral or weakly acidic solutions and on oxide-free surfaces, phase I dissolution is linked with the formation of a porous adherent film of cuprous oxide

$$Cu + \tfrac{1}{2}H_2O = \tfrac{1}{2}Cu_2O + H^+ + e$$

(The Pourbaix diagram for these conditions indicates that the product of dissolution is more likely to be cuprous oxide than divalent copper.)

The onset of phase II dissolution is associated with the formation of a thick film of basic cupric chlorate which is loosely adherent

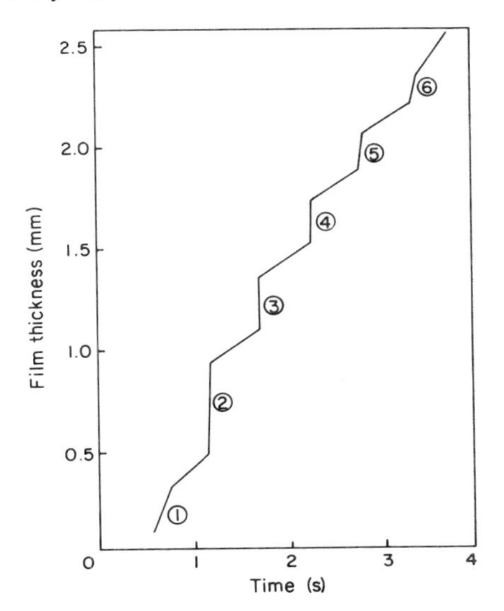

Fig. 4.12 Growth of film thickness during dissolution of copper in 3N $NaClO_3$ ($J = 3{\cdot}12$ A/cm^2); numbers indicate first, second, etc. potential peaks after start of ECM (after Cooper et al. [21])

to the anode surface. This process may also be related to the oxidation of Cu_2O by the chlorate ion:

$$\tfrac{1}{2}Cu_2O + \tfrac{1}{6}ClO_3^- + H^+ = Cu^{++} + \tfrac{1}{6}Cl^- + OH^-$$

Since a reaction of this type is much influenced by hydrogen ion concentration and temperature within the pores of the original Cu_2O film, the period of oscillation should be expected to increase with bulk hydrogen ion concentration, and indeed was observed in this work. (A decrease in pH should decrease the rate of growth of the films of basic salts.) The observed increase in period with flow-rate can also be explained from a greater rate of dissolution of soluble reaction products.

The potential peaks may be caused by reduced porosity of the surface film. This reduction is accelerated by the continual addition of solid products of dissolution. The dissipation of electrical power within the pore volume may then cause local vaporisation of the electrolyte. If this happens, the film should rupture and be removed. This explanation is again consistent with measurements of the highest rates of temperature increase at the anode at the onset of film breakdown.

Recently, studies have been reported on similar periodic phenomena encountered with low-carbon steel in 5M $NaClO_3$ [22].

4.5.8 *Variation of type of finish over a machined surface*

The transition from the active to transpassive modes of dissolution has been seen to be greatly influenced by the electrolyte flow-rate and current density. Variation in local conditions of flow-rate and current density can cause a consequent variation in finish over the anode surface.

For instance, suppose that upstream a metal is undergoing polishing, but that downstream some disturbance causes a local increase in

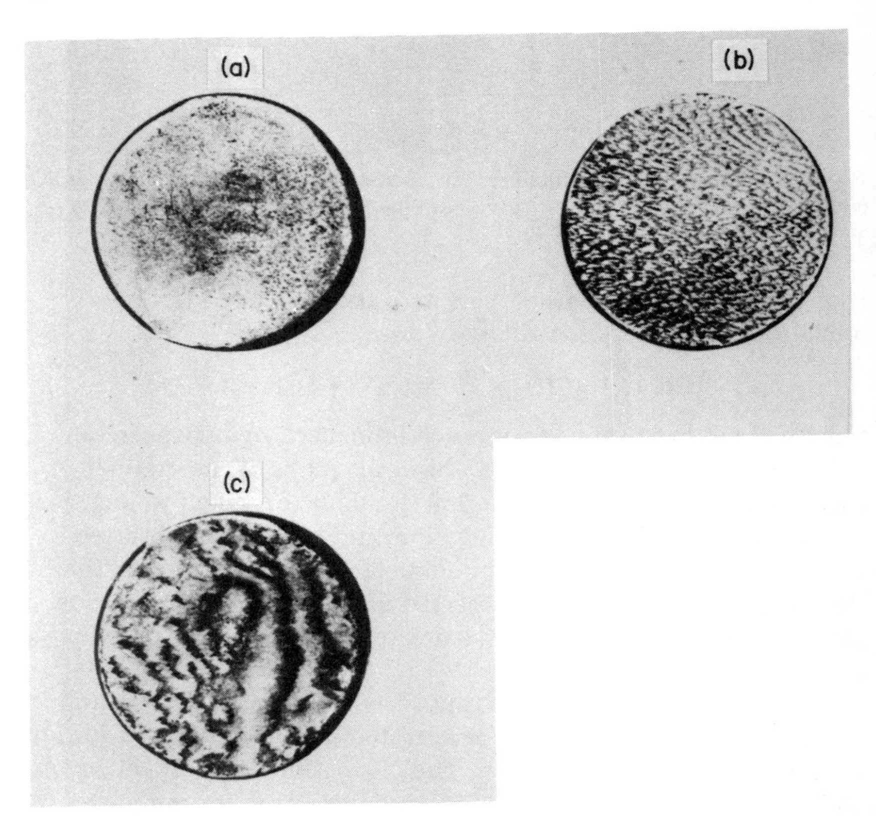

Fig. 4.13 (a) Matt, etched finish, representative of that found with nickel machined in 20% (w/w) NaCl at 35 A/cm^2, 12 V, 0·55 x 10^{-3} m^{3}s, and with Monel alloy, machined at 36 A/cm^2; (b) polished striations (Nickel machined at 58 A/cm^2, 0·36 x 10^{-3} m^3/s, 12 V); (c) selective dissolution (Monel machined at 12 A/cm^2)

electrolyte velocity. Over the latter region, an etched surface could be obtained. Alternatively, local variations in velocity could result in intermittent changes from etching to polishing or partial passivation conditions. The extent of surface striations so produced can be diminished by a change in flow-rate so that ECM is controlled by a single process over the whole surface.

Surface variations of this nature are often encountered. In one study [5] on nickel machined in HCl solution an etched surface finish without flow marks was obtained at 7·8 A/cm^2 and 2 V, the Reynolds number being 4700 (flow-rate 0·06 x 10^{-3} m^3/s; gap width 0·25 mm). However, at 18·6 A/cm^2 and 4 V the etched surface also bore polished patches with flow streamlines. In this case, ECM must have occurred in the transition region between the active and polishing modes of dissolution. For higher current densities, 34 and 64 A/cm^2, and 8 V and 12 V respectively, polished finishes were obtained, although flow marks were still also evident.

In Section 4.5.4, partial passivation was described as a characteristic of some carbon steels which had undergone various heat treatments [16]. In that work, some specimens were also found to have a variation in finish over their surface. For example, 0·78% C quenched and tempered steel had the marks of partial passivation at an electrolyte velocity of 16·5 m/s but this passivation broke down as the velocity was increased to 27 and 45 m/s, and the surface then carried polished striations. These are thought to be due to local variations in flow over the anode surface causing fluctuations between conditions favourable to partial passivity and to polishing. In other cases, the fluctuations led to variations between etching and partial passivation, and etching and polishing.

These results are similar to those found in ECM experiments on nickel in 20% (w/w) NaCl. Figure 4.13 shows a matt etched finish obtained at 35 A/cm^2, whilst at 58 A/cm^2 the finish is polished but 'broken'.

4.6 Surface films and dimensional control

The studies described in the preceding section emphasise the close link which exists between surface finish and anodic films in ECM. The link also extends to another important aspect called *dimensional control.*

4.6.1 Dimensional control

This feature of an electrolyte can be explained by reference to Fig. 4.14 which shows an electrode configuration consisting of an initially plane anode ABCD and a cathode of simple shape PQRSTU.

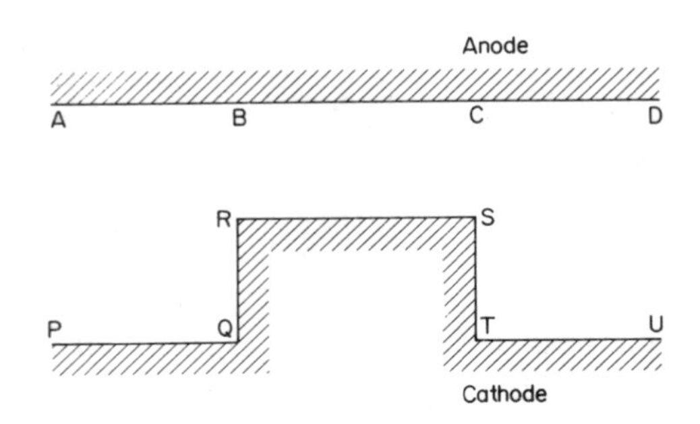

Fig. 4.14 Demonstration of dimensional control; plane anode and cathode of simple shape

Suppose that only the passage of current between RS and BC is required. In addition, however, stray current will pass from RS to the regions BA and CD, its density becoming less in the directions B to A and C to D. If surface polishing is sought along BC, the conditions of lower local current density along BA and CD may lead to etching and pitting in those regions.

The extent to which current strays in this fashion is related to the *throwing power* of the electrolyte. The corresponding problem is well known in electrodeposition. In that process, the ability of an electrolyte to provide a deposit of uniform thickness over an electrode is measured in terms of its throwing power; 100% throwing power produces a deposit of uniform thickness. In ECM, stray current is reduced at low values of throwing power; when the latter quantity approaches $-\infty$, stray current should be eliminated.

Several methods for measurement of the throwing power of solutions under ECM conditions have been reported. Brook and Iqpal [23] have modified the Haring–Blum cell which is normally used for deposition work. They have also adapted for ECM the empirical formula employed in electrodeposition for measurements of throwing power, T:

$$T = \left(\frac{L - M}{L + M - 2}\right) \times 100 \qquad (4.11)$$

Here L is the 'linear ratio' of the distances of equipotential plane parallel anodes from the cathode; M is the 'metal distribution ratio' of the weight of metal dissolved from the nearer anode to the weight dissolved from the farther anode. In their experiments, two plane

Table 4.11 Throwing powers of $NaClO_3$ and NaCl electrolytes at 25°C (after Brook and Iqpal [23])

Electrolyte	Gap width 0·125 mm	Gap width 0·25 mm
	Throwing power	
5% $NaClO_3$	−13	7
	−21	6
		3
5% NaCl	26	19
	11	12
5% NaCl + 0·12% $K_2Cr_2O_7$	−25	
	−22	
NaCl + 1% BTZ	−17	
	−15	

anodes made of 18/8 stainless steel were placed on either side of a plane copper cathode. The linear ratio was usually maintained at 5 : 1. The smaller electrode gap width was either 0·125 mm or 0·25 mm. The current density was maintained at 2·3 A/cm^2 and the flow-rate at $0{\cdot}01 \times 10^{-3}$ m^3/s. The results presented in Table 4.11 demonstrate that $NaClO_3$ has a lower throwing power than NaCl. To test the hypothesis that the low throwing power of $NaClO_3$ is linked to passivation of the anode in local current density regions, two passivation agents, benztriazole and potassium dichromate, were added to the NaCl solution. Table 4.11 shows that with these additives the throwing power of NaCl is markedly decreased. Its dimensional control is thereby improved.

An alternative method for measuring throwing power in ECM (at current densities about 95 A/cm^2) has been proposed by Chin and Wallace [24]. Their technique consists of measuring the electrode gap width as a function of machining time and expressing the linear ratio, L, and the metal distribution ratio, M, in terms of the gap width and cutting rate respectively:

$$L = \frac{h(t)}{h(0)} \quad \text{and} \quad M = \left(\frac{\mathrm{d}h}{\mathrm{d}t}\right)_0 \bigg/ \left(\frac{\mathrm{d}h}{\mathrm{d}t}\right) \tag{4.12}$$

Here $h(t)$ is the gap width at time t, and $h(0)$ is the initial width; $\mathrm{d}h/\mathrm{d}t$ is the rate of change of gap width with time, the subscript denoting the initial rate. They assume a logarithmic relationship between M and L, $M = L^{1/A}$, where A is the logarithmic throwing index which is constant for a given electrolyte at constant values of the process variables. For plane parallel electrodes, they deduce an equation which describes the variation with time of the machining gap:

$$t = \frac{h(0)(L^{1/A} - 1)}{\left(\frac{\mathrm{d}h}{\mathrm{d}t}\right)_0 \left(1 + \frac{1}{A}\right)} \tag{4.13}$$

Measurements of gap width as a function of machining time now yield that $t \propto L^n$; e.g. $n = 1{\cdot}9$ so that $A = 1/(1{\cdot}9 - 1) = 1{\cdot}11$. For a given linear ratio, L, the metal distribution M can now be calculated from $M = L^{1/A}$. A log–log plot of M against L should now give a straight line which passes through a common point represented by the linear ratio equal to 1, and by the metal distribution ratio equal to 1.

Landolt [25] has investigated the difference in throwing power between a non-passivating (acidified 1M NaCl) and a passivating (neutral 1M $NaClO_3$) electrolyte used for the ECM of nickel. His work again demonstrates that a non-passivating electrolyte has a throwing power of almost zero whilst a passivating electrolyte takes negative values.

However, little can be gained from quantitative measurements of throwing powers apart from a comparison of the performance of different electrolytes. Since calculations of this type depend on the electrode configuration and on the process variables, they reveal little information about the processes that control this property of electrolytes.

The marked difference in throwing power between a passivating electrolyte, e.g. $NaClO_3$, and a non-passivating one, e.g. NaCl, does not seem to be due to any characteristic effect that these types of solution might have on the current distribution within the machining region. Such effects will be discussed in Chapters 5 and 6. A measure of their influence will be given there in terms of a dimensionless parameter, given by the product of the electrolyte conductivity and the slope of the voltage–current density curve, divided by a typical length, e.g. the gap width. The disparity in results between a passivating and a non-passivating electrolyte may be caused by a different dependence upon current density of the current efficiency for dissolution. In Landolt's studies of throwing power, for instance, the current efficiency for dissolution of nickel in NaCl solution was not influenced by current density, whilst that in $NaClO_3$ was strongly dependent on both current density and flow-rate. Nonetheless, much remains to be done to establish the real nature of throwing power in ECM.

4.6.2 The nature of surface films in different electrolytes

Since the quality of surface finish and dimensional control are closely associated with surface films, this section is devoted to a study of the nature of the anodic films formed in different electrolytes.

Much consideration has been given to the superior ECM results obtained with $NaClO_3$ solution compared with those found with NaCl electrolyte. By means of the polarisation curves shown in

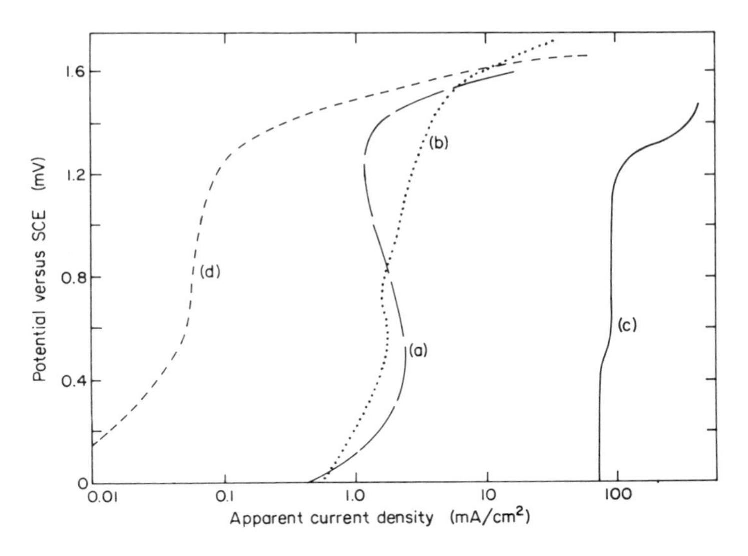

Fig. 4.15 Polarisation curves for iron; (a) $NaClO_3$; (b) $NaNO_3$; (c) NaCl; (d) $Na_2Cr_2O_7$ (after Hoare et al. [26])

Fig. 4.15, the performance of iron in those electrolytes and in $NaNO_3$ solution is first discussed [26].

In relating these polarisation curves to ECM conditions, we shall look for the range of anode potentials over which dissolution is possible, bearing in mind that in ECM the *IR* drop between the electrodes takes large values for small inter-electrode gap widths. If the *IR* drop becomes sufficiently large, the potential at the anode may decrease so much that dissolution ceases (see Chapter 3).

(a) $NaClO_3$ electrolyte

For $NaClO_3$, active dissolution occurs at the lowest potentials; the iron dissolves as Fe^{++} ions which can be oxidised by dissolved oxygen to Fe^{+++} ions. As the potential is increased to more noble values, a critical current value is reached at which the anode surface becomes covered with a film. Further studies [27, 28] have shown that this film is γ-Fe_2O_3 and that its thickness is about 10^{-7} m. As the film forms over the active sites on the metal, dissolution is inhibited by the onset of passivation with corresponding reduction in current. On further increases in the potential, the film is broken down, and the transpassive region of dissolution is reached with an equivalent increase in current. ECM is assumed to occur in this region.

Figure 4.15 shows that with $NaClO_3$ solution the transition from passive to transpassive conditions is abrupt. Accordingly, the potential range between dissolution and non-dissolution is very narrow. Hence, in a shaping operation with $NaClO_3$ electrolyte, ECM should occur mainly in the region of smallest gap, with little ECM taking place outside that region. Good dimensional control is then achieved.

(b) $NaNO_3$ electrolyte

For $NaNO_3$, the transition from passivity to transpassivity is less sharp than that for $NaClO_3$. ECM over a wider region on the anode can therefore be expected. Since the transpassive region also lies at a more noble potential for $NaNO_3$ than for $NaClO_3$, the machining rate at the same operating conditions for $NaNO_3$ solution should be less than that for $NaClO_3$. This observation has been confirmed by ECM experiments in which the current efficiencies for metal removal and oxygen generation are about 12% and 85% respectively [29]. These processes appear to be influenced by the formation of a film of Fe_3O_4 on the anode surface. The film also seems to be electronically conducting and to favour a chemical reaction other than metal dissolution.

Since the thickness of this film is not noticeably decreased in the transpassive region, the surface finish is not likely to have the brightness expected when a thinner compact film, commonly associated with electropolishing, is present. ECM experiments again bear out this suggestion; reported typical surface roughnesses in 3M $NaNO_3$ are about 0·65 to 1 μm RMS.

(*c*) *NaCl electrolyte*

For NaCl, the current density in the polarisation curve is seen to be much higher than that for the other electrolytes. Dissolution is then possible at even the lowest potentials. Consequently, in ECM, dissolution will still occur even at very wide gaps, resulting in considerable stray machining and poor dimensional control. These machining conditions are probably due to an iron salt on the anode surface, whose presence prevents the metal changing from its active state. The film appears to be a thick, unprotective layer of porous salts, e.g. $Fe(OH)_2$ [28].

(*d*) $Na_2Cr_2O_7$ *electrolyte*

Since dimensional control with $NaNO_3$ electrolyte is better than that with NaCl, but poorer than $NaClO_3$, it has been suggested that the presence of an anion which contains oxygen may be necessary if stray current is to be diminished. This premise has led to a survey of chromate electrolytes, since a protective film of γ-Fe_2O_3 is known to form on iron in these electrolytes. From Fig. 4.15, the polarisation curve for $Na_2Cr_2O_7$ solution indicates the presence of a strongly passivating film. Although this film is removed abruptly in the transpassive region, the potentials then are much higher than for $NaClO_3$. In ECM, electrical sparking becomes likely. Indeed, poor ECM results have been obtained with $Na_2Cr_2O_7$ solutions, and with other passivating electrolytes, Na_2CO_3 and Na_3PO_4, no machining has been possible. Polarisation and ECM experiments with steel in $NaClO_4$ and Na_2SO_4 electrolytes and mixtures of them have also shown that these electrolytes are inferior to $NaClO_3$ [30]. $NaClO_3$ electrolytes appear to be unique in ECM in that the transition from the passive to the transpassive region is a sharp one; also, the latter regions occupy potentials which are the least noble.

On this basis, it has been suggested that good dimensional control requires (i) the formation of an anodic passivation film, and (ii) the occurrence over a narrow band of potentials of the transition from passivity to transpassivity [27]. This view, however, is not universally held. Landolt contends that the width of the passive potential region in a potentiostat current–potential curve should bear little influence on current distribution in ECM [25]. The basis for this claim is that the polarisation parameter, which dictates the extent to which current distribution in ECM is affected by an electrolyte's throwing power, depends only on the slope of the current–voltage curve, and not on actual values of the anode potential.

Although the merits of $NaClO_3$ are undoubted, similar dimensional control has also been claimed to be possible with simple NaCl solutions [31, 32]. The reasoning behind this claim can be clarified from the anode potential-anion concentration diagram, Fig. 4.16. This diagram shows the likely surface finish produced for particular conditions of potential and concentration. Suppose that machining is performed in a region where the conditions are given by A. The surface finish then achieved should be bright and polished. In a neighbouring zone which is at lower potentials, B, passivity will occur. If chloride ions are present, they are likely to attack the passive film causing pits to

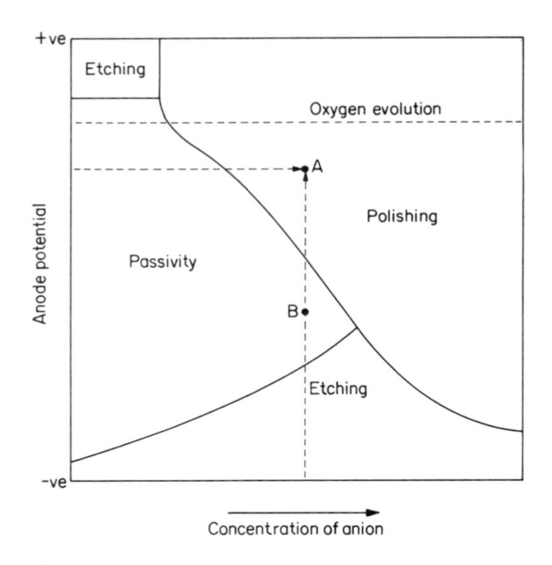

Fig. 4.16 Influence of anion concentration and anode potential upon surface finish (after Boden and Evans [32])

form. On the other hand, the presence of chlorate ions should produce only a mild attack. Clearly, the ability of the passive film to resist anion attack is critical if stray current is to be diminished. But even if a passive film is formed, its efficacy can be spoiled by severe pitting in the passive (stray current) region; also the problem of choice of a suitable additive becomes aggravated by the need to reduce the depth and number of pits in the stray-current zone without affecting machining in the main region, A.

Suitable solutions appear to be those which form insoluble salts, e.g. with nickel anodes, a combination of NaCl with Na_2CO_3 forms insoluble nickel carbonate. Combinations of solutions which produce soluble salts behave like NaCl, e.g. NaCl with Na_2SO_4. For nickel machined at a flow-rate of 30 m/s and a maximum current density of 15·5 A/cm^2 in 15% (w/w) NaCl containing 2·5% (w/w) Na_2CO_3, the amount of pitting was found to be considerably less than that of the simple 15% NaCl solution; indeed, the performance of the former solution almost matched that of a 40% $NaClO_3$ electrolyte. In contrast, a solution of 15% NaCl + 1% Na_2SO_4 gave only a slight reduction in pitting owing to the formation of a relatively soluble nickel sulphate salt.

The improvement gained with insoluble nickel salts appears to be due to the precipitation of these salts around the pits in the stray current region. But success also depends critically on the quantity of anion addition to the basic sodium chloride electrolyte. For instance, in the ECM experiment above, precipitation only occurred in the stray current zone provided that the percentage of sodium carbonate in the 15% NaCl solution was below 2·5. Brightly machined surfaces with high current efficiencies could then be achieved in the main machining zone. But at higher concentrations of Na_2CO_3, precipitation also occurred in the main zone with the onset of passivation, that is, coverage of the surface with insoluble nickel carbonate and a reduction in machining rate.

These observations indicate that the ECM of iron or steel in $NaClO_3$ is quite different from that of nickel in this electrolyte. This has been confirmed in tests on nickel machined in a solution of 350 g/l $NaClO_3$ [33]. Some results are summarised in Table 4.12. They show that the current efficiency for nickel removal does not increase as rapidly with current density as it does for mild steel, nor does it achieve as high values. It was also found that, under identical operating conditions, the maximum machining rate for

Table 4.12 Current efficiency values for steel (type 1020-U.S.) and nickel (99·5% pure) machined in 350 g/l $NaClO_3$ solution (after LaBoda et al. [33])

Metal	Current density (A/cm^2)	Current efficiency (metal removal) (%)	Current efficiency (O_2 generation) (%)
Steel	40	0·3	–
	80	22	78
	150	62	40
	300	86	18
Nickel	40	8	–
	80	15	79
	150	58	–
	300	67	40

nickel was about one-half of that for mild steel (0·07 mm/s and 0·13 mm/s respectively). Corresponding roughness measurements also showed a contrast: 0·26 to 0·39 μm for nickel against 0·026 to 0·052 μm for steel. These results are consistent with polarisation experiments which show that, with increasing potentials, the anodic film on nickel thickens, unlike the film on mild steel which under similar conditions becomes uniformly thinner and favourable towards electropolishing conditions.

This work again establishes that the performance of a metal in ECM varies from one electrolyte to another and depends greatly upon the nature of the anodic films which are formed and upon their interaction with the anions in the electrolyte.

4.7 Surface finish: some macroscopic effects

4.7.1 Selective dissolution with alloys

When a metal has constituents whose electrode potentials are different, those with the lowest potential will first be dissolved. At low current densities, the result at the surface of the metal is differential dissolution of the constituents, observable as a variation in surface finish. At higher current densities, the effect of differences in electrode potentials of constituents is reduced, since the applied potential difference at the anode is then higher, and the potentials required for dissolution of the constituents are more easily achieved.

A similar effect can be found with a metal whose surface includes a grain of different electrode potential from that of the surrounding material. If the electrode potential of the grain material is greater, then its dissolution will not occur until the potential at the anode surface reaches its electrode potential. During that time, the surrounding material will be machined, leaving the grain protruding at the surface. On the other hand, if the grain has a lower electrode potential, its dissolution will proceed first, leaving a recess on the surface. This effect is often observed as grain boundary attack. Owing to the traction forces created by the high rates of electrolyte flow in ECM, grains can even be dragged from the metal surface; increased surface roughness is then observed.

Photographs of Monel alloy machined in 20% (w/w) NaCl solution are presented in Fig. 4.13 to illustrate selective dissolution. At 12 A/cm^2, the copper (dark lines) is separated from the nickel constituents (light lines). At the higher current density, 36 A/cm^2, the effect is less obvious, the surface being smoother and more uniform.

Attack along the grain boundaries of Nimonic 80A dissolved in 0·1M to 4·3M NaCl has been the subject of a detailed study by Evans and Boden [34]. They found that the severity of grain boundary attack could be lessened by a decrease in electrolyte flow-rate, from about 30 to 6 m/s, the maximum current density being about 1·5 mA/cm^2. They postulate that, because of the high rates of dissolution along the grain boundary regions, conditions favourable to polishing should first be achieved. But for the high rates of metal removal to be maintained there, the process of dissolution becomes greatly dependent on the rate of arrival of water molecules. At lower electrolyte flow-rates, this rate of arrival will be decreased. Salt passivation then arises, and with it a reduction in the amount of grain boundary attack.

4.7.2 Flow separation

An anode surface, which initially is irregular, may also cause flow separation between the 'hills' and the 'valleys' on its surface. At this stage, a short discussion on the possible effects of flow separation is useful, particularly since it reinforces the explanations given in

Section 4.5.8 for the variations of finish over the surface, which are so often observed.

The basis of the argument is that, over a flat anode, the diffusion layer can be assumed to be adjacent to the surface. Metal removal is controlled by the usual diffusion-migration mechanism, discussed

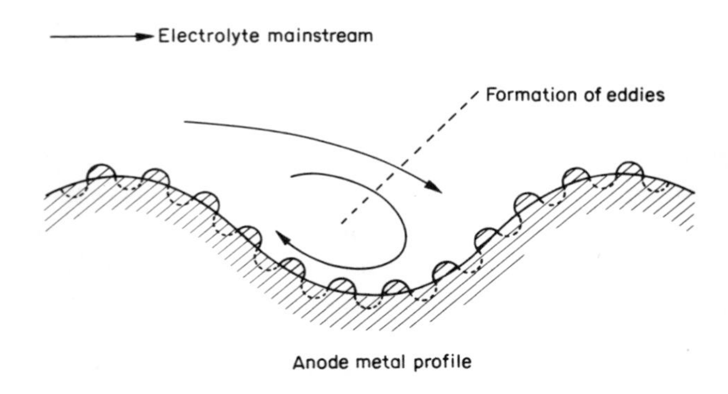

Fig. 4.17 Formation of eddies in 'valley' of anode metal

in Chapter 3, and because of the high electrolyte velocity, concentration polarisation can be assumed to be negligible. But the flow over an irregular anode could be quite different. Around the hills, the flow still adheres to the metal surface. Metal removal continues to be governed by the processes outlined above. On the dissolution of both large and small scale irregularities, the result should be a polished surface in that region. In the valleys, a different situation arises, as suggested in Fig. 4.17. Here the flow separates from the metal surface, causing a rotating eddy in the region between the mainstream of the electrolyte and the metal surface in which a large uniform concentration of metal ions may be built up. This may cause a high concentration overpotential in that region, making difficult the local removal of the smallest irregularities. Polishing is not then achieved, and a matt, etched finish can be expected. This could explain the type of 'broken' polished finish, observed by Cuthbertson and Turner [6]. If the pre-machining preparation of the anode surface, carried out by them to remove an oxide film, also involved some smoothing of the surface, this, of course, would also reduce the initial heights of surface irregularities, causing a more uniform flow pattern. This, in turn, would reduce the probability of eddies, and overall polishing would again be more likely.

Freer, Hanley, and MacLellan [16] have also suggested that observed unevenness and striations, superimposed on surfaces after ECM, could be caused by local variations in flow. If the flow conditions lead to electropolishing (diffusion control) over the hills, the lower rates of flow in the valleys may cause etching (activation control) in those regions. Local variations in rate of flow may also account for surface unevenness due to fluctuations between conditions favourable to partial passivation and either activation or diffusion control. Partial passivation will lead to a locally raised region (due to reduction in machining rate) protruding over the regions where activation or diffusion mechanisms have caused the rate of dissolution to be greater. Sufficiently high rates of flow can reduce the extent of all these effects.

Non-uniform flow can also arise from cavitation phenomena. Even if their effect is not sufficiently great to cause a short-circuit, the variation in the flow pattern usually becomes apparent by striations developed on the anode surface.

4.7.3 Hydrogen gas

As the electrolyte flows downstream within the machining gap, it collects in solution an increasing amount of the hydrogen gas generated at the cathode. The presence of the gas–electrolyte mixture has diverse effects on the ECM process, many of which will be discussed in Chapter 5. But a principal effect can be appropriately introduced here, namely, a decrease in effective conductivity of the solution. This decrease becomes more marked in the downstream direction, and has a consequential influence on surface finish.

Further observations from the work above [32] become relevant. In that study, plane parallel electrodes, spaced 0·25 mm apart, were used; the flow-rate was 30 m/s and the current density 15·5 A/cm^2. Polishing of the nickel anode was sought over the central machining region. At the same time, attempts were made to reduce the depth and number of pits in the passive, stray current zone situated upstream and downstream. The extent of stray current attack was found, in fact, to be diminished by a reduction in the concentration of the NaCl solution, e.g. from 20 to 13·5%. But in the downstream, stray current zone, a further decrease in pitting was achieved. This was attributed to a decrease in the effective conductivity of the solution by the accumulation of hydrogen gas bubbles. But the addition of small quantities of Na_2SO_4 (1%) and Na_2CO_3 (2·5%)

to a main 15% (w/w) NaCl electrolyte led to increased pitting in these areas, mainly because these mixtures of electrolytes have a higher conductivity and are therefore less affected by the hydrogen gas.

The specimens used in this study were ground and mechanically polished so that they had a perfectly flat surface before ECM. In practice, however, some previous machining or working operation could make the anode surface slightly irregular before its treatment by ECM. For a surface of this initial form, a theoretical investigation is available which deals with the effects of hydrogen gas on the eventual finish achieved by ECM [35]. The bases of that study are considered in detail in Chapter 6, since they also cover the problem of anodic shaping in its simplest form, anodic smoothing. But those features relevant to the present study can be introduced here.

Suppose now that the anode surface is not initially flat, but that it consists of a series of irregularities of widely varying horizontal and vertical dimensions. Smoothing is achieved by the non-uniform dissolution of the anode surface, the peaks being dissolved faster than the valleys. The removal of the smallest irregularities constitutes 'polishing' or 'brightening'.

The theory of anodic smoothing (Chapter 6) predicts that for a flat surfaced cathode and an anode with a sinusoidal irregularity of wavelength λ ($= 2\pi/k$, where k is the wave number) and amplitude ϵ which is small compared with the average gap width p, the irregularity is reduced at a rate given by

$$\epsilon(t) = \epsilon(0) \exp\left(-\frac{MVk \coth kp}{p} t\right) \tag{4.14}$$

where $\epsilon(0)$ is the initial amplitude, M a machining parameter ($= e_a \kappa_e / \rho_a$), e_a being the electrochemical equivalent of the metal and ρ_a its density, and κ_e the electrolyte conductivity, V the applied potential difference, and t the machining time. This formula is valid provided that the gap width is maintained at its steady-state value,

$$p_e = \frac{MV}{f} \tag{4.15}$$

determined from the equation, proved in Chapter 6,

$$\frac{dp}{dt} = \frac{MV}{p} - f \tag{4.16}$$

where f is the cathode feed velocity.

The result (4.14) demonstrates that short-wavelength irregularities are removed more rapidly than those of longer wavelength. We shall see in Chapter 6 that Fourier analysis may be used to extend these results to describe the behaviour of an irregularity of arbitrary profile. Clearly, sharp-profile irregularities are removed more rapidly than 'bluff' ones, and the irregularities are reduced to a sinusoidal form of basic wavelength. Local polishing, as distinct from long-wavelength smoothing, occurs comparatively quickly; an overall polished surface may be obtained with low amplitude, long-wavelength irregularities. This difficulty is enhanced by the increase of the machining rate (i.e. increase of current density, or of the factor MV in Equation (4.14)).

The theory also applies to cases where the electrodes may be considered to be locally plane and parallel, i.e. if the radius of curvature of the electrode configuration is large compared with the gap width.

Variations in the surface finish due to the electrolyte flow may be explained further in terms of changes in the effective conductivity of the electrolyte. As we shall see in Chapter 5, the conductivity will be affected by factors such as electrolyte heating and hydrogen gas bubble generation. Joule heating will increase the temperature and so increase the conductivity, whilst hydrogen gas will reduce it by decreasing the volume of electrolyte available for conduction. Both effects are cumulative in the downstream direction of the electrolyte flow. Under steady-state machining conditions, these changes in conductivity are compensated by a corresponding change in the inter-electrode gap width, so that the steady-state current density does not vary locally along the electrode length. However, if such equilibrium is not achieved in the machining time required for electrochemical smoothing to a certain degree, then local variations in conductivity along the electrode length, and hence in current density and in surface finish, might therefore be expected. The following analysis has been proposed to explain these related effects of gas and flow on surface finish.

It will be recalled that Equation (4.14) for the removal of the irregularities involves a machining parameter, M, which depends directly on conductivity. If, in this simplified analysis, overpotential effects are excluded, conductivity is the only factor in the equations for smoothing which is dependent on conditions in the electrolyte. To examine its effects, we replace κ_e in Equation (4.14) by κ_m, a mean effective conductivity.

For simplicity, the gap is assumed to remain constant at its mean value p, and variations in current density and conductivity along the electrode length are sought (cf. the actual, steady-state situation where the current density is constant and the gap width varies). The mean conductivity κ_m along the gap may be expressed in terms of the conductivities in the electrolyte zone free of bubbles and the bubble layer:

$$\kappa_m = \kappa_e \exp\left(-\frac{1{\cdot}5\ \kappa_e H_g V}{Up^2}x\right) \quad (4.17)$$

where κ_e is the electrolyte conductivity in the gap zone which is free of hydrogen bubbles, H_g is the volume of hydrogen produced per coulomb, U is the electrolyte velocity, assumed constant, and x is the distance downstream. The derivation of this equation is rather lengthy; it depends on the experimentally observed results that the bubble layer thickness increases linearly along the gap [36], and that for a two-phase mixture the ratio of the conductivity in the bubble zone to the conductivity outside the zone is equal to $(1 + \alpha)^{-1{\cdot}5}$, where α is the void fraction [37]. Nonetheless, the full analysis is readily available [35].

Since the machining parameter M and the conductivity have the same dependence on distance x, then qualitatively the modified expressions for κ_m and M mean that their values decrease in the

Table 4.13 Effects of electrolyte velocity, U, and distance along the electrode, x, on the ratio $\epsilon(t)/\epsilon(0)$

Long wavelengths

	x = 10 mm			x = 50 mm			x = 100 mm		
U (m/s)	1	10	20	1	10	20	1	10	20
$\epsilon(t)/\epsilon(0)$	0·85	0·56	0·53	1	0·72	0·62	1	0·85	0·72

Short wavelengths

	x = 10 mm			x = 50 mm			x = 100 mm		
U (m/s)	1	10	20	1	10	20	1	10	20
$\epsilon(t)/\epsilon(0)$	0·19	0	0	1	0·04	0·01	1	0·19	0·04

downstream direction, or if the electrolyte velocity is decreased. If these reduced values for M are substituted in the earlier equations for $\epsilon(t)$, the conclusions are that surface roughness increases in the direction of flow and that the overall roughness decreases if the electrolyte velocity is increased.

Results of some calculations are presented in Table 4.13. They have been obtained by the substitution in Equation (4.17) of the typical values, $\kappa_e = 0{\cdot}2\ \text{ohm}^{-1}\ \text{cm}^{-1}$, $H_g = 0{\cdot}12 \times 10^{-6}\ \text{m}^3\ \text{A}^{-1}\ \text{s}^{-1}$ $p = 0{\cdot}5$ mm, $U = 1, 10, 20$ m/s, $x = 10, 50, 10^2$ mm, $V = 10$ V, $t = 10$ s. From the calculated values of κ_m, modified values for M ($= 8{\cdot}5 \times 10^{-5}$ κ_m) have been found. These values have been used in Equation (4.14) for $\epsilon(t)/\epsilon(0)$, in which a value $k = 20\ \text{mm}^{-1}$ has been taken. Note that in Table 4.13 two cases have been considered: one in which the irregularities are of long wavelength ($kp \ll 1$), and another in which the wavelengths are short ($kp \gg 1$). For the former case, Equation (4.14) reduces to

$$\frac{\epsilon(t)}{\epsilon(0)} = \exp\left(-\frac{MV}{p^2}t\right) \tag{4.18}$$

whilst, for the latter condition,

$$\frac{\epsilon(t)}{\epsilon(0)} = \exp\left(-\frac{MVk}{p}t\right) \tag{4.19}$$

Experimental results of this type have been obtained in work on Nimonic 80A machined in saturated NaCl solution [6]. In that study, specimens were machined in the same condition in which they were received, i.e. without their surfaces being mechanically treated before ECM. At a current density of 31 A/cm^2 and a flow-rate of 12 m/s, the surface roughness increased markedly in the downstream direction. An increase in velocity to 28·5 m/s had the effect of improving the quality of finish.

4.7.4 *Effects of overpotential*

The analysis in Chapter 6 upon which the results in Section 4.7.3 are based is of further interest in that it provides information about the effects of overpotential upon surface finish. Again, for brevity, only the main results from that chapter are presented here. As before, the cathode is considered to have a flat surface whilst the anode surface is initially irregular.

In Chapter 6, the basic theory is extended to include an arbitrary, current density-dependent overpotential. In that study this overpotential is initially supposed to exist only at the cathode. The overpotential is also assumed to cause a small departure in the value of the current density at the anode, J, from the mean current density, $\bar{J}$, say. The overpotential function, written here as $f(J)$, can then be expanded as a Taylor series:

$$f(J) = \alpha + \beta(J - \bar{J}) \tag{4.20}$$

where

$$\alpha = f(\bar{J}) \tag{4.21}$$

and

$$\beta = \left.\frac{\partial f}{\partial J}\right|_{J=\bar{J}} \tag{4.22}$$

Then, as shown in Chapter 6, Equation (4.14) takes the modified form

$$\epsilon(t) = \epsilon(0) \exp\left[-\omega M\left(\frac{V-\alpha}{p}\right)kt \coth kp\right] \tag{4.23}$$

where

$$\omega = 1 - \beta\kappa_e k \operatorname{cosech} kp \operatorname{sech} kp(1 + \beta\kappa_e \coth kp)^{-1} \tag{4.24}$$

Note that the quantity $\beta\kappa_e$ has the dimension of length. It is equivalent to the electrodeposition polarisation parameter, discussed by Wagner [38], which is given by the product of the electrolyte conductivity and the slope of the voltage–current density curve of the electrolytic cell.

Two particular cases can now be considered. For short-wavelength irregularities, $kp \gg 1$; ω is negligibly different from unity. The dissolution of such irregularities is unaffected by overpotentials. For longer wavelengths,

$$\omega \simeq \left(1 + \frac{\beta\kappa_e}{p}\right)^{-1} \tag{4.25}$$

That is, for large values of the dimensionless parameter, $\beta\kappa_e/p$, the rate of dissolution is zero. The overpotential also has an effect on the equilibrium gap, the expression for which becomes

$$p = \frac{M(V-\alpha)}{f} \tag{4.26}$$

The cathode overpotential, therefore, has the following principal effects:

(i) decrease of the mean machining rate;
(ii) decrease of the removal rate of surface irregularities of long wavelength.

An analogous discussion of anode overpotentials shows that the behaviour of long-wavelength irregularities is similar to the cathode overpotential case; for very high values of the overpotential parameter $\beta\kappa_e k$, however, the reduction rate of all irregularities is very small, irrespective of the wavelength. Smoothing is then very difficult, the initial, irregular anode profile being 'frozen' as machining proceeds.

4.8 Use of molten salt electrolytes

The electrolytes studied so far have been restricted to aqueous types. It will have become obvious, too, that the rates of reactions in ECM can usefully be increased by operation at high electrolyte conductivities and temperatures. Because aqueous electrolytes have their limitations (e.g. those due to their boiling temperatures) some attention has been paid to the possible use of molten salt electrolytes in ECM. These solutions have the attraction that their values of density and viscosity are similar to those of aqueous electrolytes, whilst they have electrical conductivities which are often greater by more than an order of magnitude. For example, molten NaOH has a conductivity of 2·44 ohm^{-1} cm^{-1}, a density of 1·77 g/cm^3, and a viscosity of 3·79 cP at 630°C. As a molten electrolyte, NaOH has been used for the ECM of tungsten carbide [39]. Current densities as high as 186 A/cm^2 were achieved, in comparison with a maximum 49 A/cm^2 possible with aqueous NaCl under the same experimental conditions. The metal removal rates were correspondingly higher and a smooth uniform surface finish was obtained. This electrolyte in its molten form has been thought to be a better candidate than NaCl because of its lower melting point (about 320°C compared with 800°C for NaCl).

Lovering has studied the polarisation behaviour of nickel, Nimonic, and titanium alloys in molten KNO_3 electrolyte at 350°C [40]. He reports that *cathodic* polarisation leads to the dissolution of these metals, the apparent current efficiency being less than 10%. The anodic polarisation of titanium leads to the development of a thick, electronically conducting film on the metal surface, with O_2 gas being evolved for applied potential differences up to 30 V. Oxygen gas is also invariably generated with Nimonics. On the other hand, anodically polarised nickel dissolves easily at low voltages (0·5 to 1 V) and without gas evolution.

Further experiments with LiCl–KCl electrolytes at 400°C have again confirmed that nickel based alloys can be dissolved at low anodic potentials. Apparent current efficiencies are often in excess of 100%, although these values can be attributed to the erosion of intermetallic particles. At higher potentials, chlorine gas evolution occurs in addition to metal removal. But LiCl–KCl electrolyte is not suitable for titanium and its alloys. In this electrolyte these metals dissolve spontaneously with simultaneous gas evolution.

A disadvantage of molten salt electrolytes is that they are difficult to handle. At present they are unlikely to replace aqueous solutions in ECM.

4.9 Effects of ECM on the mechanical properties of metals

Since conventional machining operations are known to modify the surface structure and mechanical properties of metals, the likelihood of corresponding effects occurring from ECM requires study.

First, no cases are known of hydrogen embrittlement of the anode surface, which, of course, can affect the ductility of the metal, and which is the cause of metal failure in some structures. Other tests have confirmed that ECM does not affect the ductility, yield strength, ultimate tensile strength, or microhardness of metals [41]. These results appear to be applicable to most metals encountered in ECM. In fact, for some metals, e.g. beryllium and tungsten, ECM is claimed to be able to improve these mechanical properties by the removal of surface layers, which have been damaged by conventional machining. (But for these metals, the choice of a suitable ECM electrolyte is another problem, since surface defects produced by ECM may be a source of reduced fatigue strength).

With many metals, mechanical machining raises the fatigue strength by the introduction of compressive stresses to its surface layers. If, after a mechanical operation, a metal undergoes ECM, a reduction in compressive stress should arise through the electrochemical removal of the surface layers. In Figs. 4.18 and 4.19 confirmatory experimental evidence is presented for Nimonic 80A

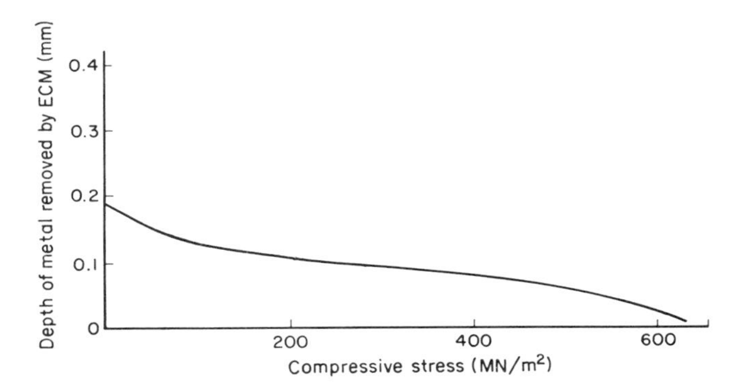

Fig. 4.18 Effect of depth of metal removed upon surface compressive stress (after Evans et al. [42])

machined in NaCl and $NaClO_3$ solutions [42]. Compressive stresses were found to exist to about 0·2 mm depth (Fig. 4.18). The electrochemical removal of this depth of metal has the effect of reducing the fatigue life to a constant value (Fig. 4.19).

Note that a stress-free surface so produced is one from which the true fatigue strength of the metal can be measured. If, after mechanical machining, the fatigue strength of a metal is reduced by an ECM operation, the required strength can be restored by further mechanical treatment. A sequence of such practices for 403 stainless steel is shown in Table 4.14 [43]. Here, the fatigue strength produced

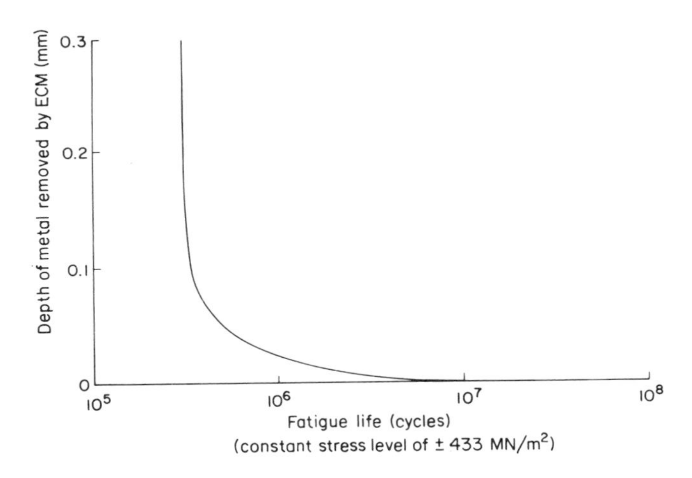

Fig. 4.19 Effect of depth of metal removed upon fatigue life (after Evans et al. [42])

Table 4.14 Effects on various surface finishing operations on fatigue life of 403 stainless steel (after Zimmel [43])

Sequence of surface treatments	Fatigue strength at $3{\cdot}1 \times 10^7$ cycles (MN/m^2)
Mechanical polishing to 0·4 μm surface finish	470
ECM to 0·76–1 μm finish	350
ECM followed by vapour blasting to 0·38–0·51 μm finish	465
ECM followed by glass-bead blast to 0·38–0·51 μm finish	510

by mechanical polishing has been lowered by ECM. After ECM, treatments by vapour blasting and by glass-bead blasting restored the fatigue strength of the metal to its former levels.

The various surface finishes produced by ECM may themselves be sources of reduction in fatigue properties. An examination of effects of this type has been carried out on Nimonic 80A. The surface finishes studied and their conditions of preparation by ECM are summarised in Table 4.15. Micro-examination of brightened

Table 4.15 Fatigue lives produced by different surface finishes at a stress of ±386 MN/m^2 (after Evans et al. [42])

Surface finish	Fatigue life (cycles)
Electrochemically polished	$4{\cdot}9 \times 10^5$
Etched	$4{\cdot}4 \times 10^5$
Integranular attack	$4{\cdot}25 \times 10^5$
Hemispherical pits	$3{\cdot}3 \times 10^5$

surfaces and brightened surfaces with grain boundary delineation showed that fatigue cracks usually originate from pits and that, owing to the electrolyte flow, these defects are more severe than those found in electropolishing. Nevertheless, little consequent reduction in fatigue life occurs. Measurements of the depths of intergranular attack showed that these were about 10^{-2} to 10^{-3} mm; again, defects of this type appear to have little effect on metal fatigue.

Related work on the fatigue properties of electrochemically machined titanium alloys has confirmed that pitting and intergranular attack reduces only slightly the fatigue life of these metals [44].

The removal of compressive stress appears to be the main reason for possible reduced fatigue life of metals undergoing ECM. Surface defects caused by the process itself seem to have only effects of secondary importance.

4.10 Effects of ECM on the properties of the electrolyte solution

In this section, we examine the extent to which continual machining affects the bulk properties of the electrolyte solution.

4.10.1 Specific gravity, conductivity, and viscosity

The metallic products of machining are thought to have negligible effect on the bulk specific gravity and conductivity of the solution [2, 45]. Confirmatory evidence is available from tests on a nickel based alloy machined in NaCl electrolyte of bulk specific gravity and specific conductivity 1·17 and 0·16 ohm^{-1} cm^{-1}, respectively, at

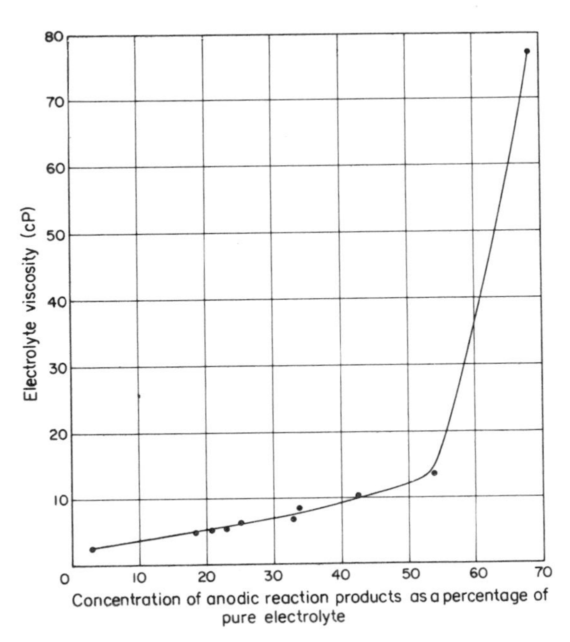

Fig. 4.20 Effect of concentration of anodic metal products on the electrolyte viscosity (after Bayer et al. [45])

24°C; in this work, the anodic debris was present in concentrations ranging from 3 to 70% (v/v) of the electrolyte solution. But the viscosity of the solution does increase with increasing quantities of the anodic products (Fig. 4.20).

4.10.2 pH

That the bulk pH of solutions can change substantially is seen also from Mao's work. For mild steel machined in 2·0M to 4·5M concentrations of NaCl, he reports pH changes from initial values about 6·0 to about 10. These final values indicated the saturation of the solutions with greenish-black coloured $Fe(OH)_2$. In $NaClO_4$ solutions of 2·0M to 4·5M concentrations, the pH changed from about 5·5 to 9. This pH change appeared to be consistent with reaction products of Fe_3O_4 in solution. For 2·0M to 4·5M $NaNO_3$ solutions, the pH values increased with ECM from about 5–6 to about 11. This substantially higher pH is attributed to the formation of ammonia. This reaction, with the formation of nitrite and hydroxylamine, is preferred to hydrogen gas evolution at the cathode in this electrolyte. For $NaClO_3$, the pH values decreased from about 7–8 to about 5–7. This decrease in pH value after ECM could have resulted from a saturated $Fe(OH)_3$ solution, characterised by its red-brown appearance.

Simple calculations have been carried out which show that, if the effect of a cathode reaction is the release of OH ions into the solution, the pH change may be sufficiently great to cause the precipitation of hydroxides of metals where the pH for precipitation exceeds that of the bulk solution [12]. The authors show that, for typical ECM conditions in which the current is 100 A, the electrolyte velocity is 36 m/s, and the machining region has dimensions 25 x 25 x 0·5 mm^3, the pH change decreases with the initial pH; in solutions whose initial pH is about unity, the change in pH is small. But if the initial pH is nearer neutral values, the pH can increase by several units. It is proposed that if a pH of about 8 is required to precipitate nickel hydroxide, a saturated salt solution, with pH 6 to 7, would seem to be unlikely to produce it. But the calculations indicate that the pH will be increased by ECM to cause the precipitation of $Ni(OH)_2$.

4.10.3 Ageing

Earlier, the throwing power of NaCl electrolyte solutions was noted to be significantly decreased by the addition of potassium dichromate.

It has been suggested that the presence, in an electrolyte, of increased quantities of chromate ions derived from the chromium present in anode metals can have the effect of reducing the throwing power of electrolyte. This effect could be responsible for the so-called 'ageing' of electrolytes through prolonged use [23].

References

1. Boden, P. J., Brook, P. A. and Evans, J. M., Paper presented at Electrochemical Engineering Symposium, The University, Newcastle upon Tyne, March 1971.
2. McGeough, J. A., Ph.D. Thesis, Glasgow University (1966).
3. Kinoshita, K., Landolt, D., Muller, R. H. and Tobias, C. W., *J. Electrochem. Soc.* (1970) **117**, No. 10, 1246.
4. Turner, T. S., Ph.D. Thesis, Nottingham University (1966).
5. Larsson, C. N., Ph.D. Thesis, C.N.A.A. (1968).
6. Cuthbertson, J. W. and Turner, T. S., *The Production Engineer* (1966) May, 270.
7. Mao, K. W., *J. Electrochem. Soc.* (1971) **118**, No. 11, 1876.
8. Konig, W. and Degenhardt, H., in *Fundamentals of Electrochemical Machining* (ed. C. L. Faust), Electrochemical Society Softbound Symposium Series, Princeton (1971) p. 63.
9. Bergsma, F., Paper presented at I.S.E.M. Conf., Vienna, October 1970.
10. Mao, K. W., *J. Electrochem. Soc.* (1971) **118**, No. 11, 1870.
11. Chikamori, K. and Ito, S., *Denki Kagaku* (1971) **39**, 493.
12. Boden, P. J. and Brook, P. A., *The Production Engineer* (1969) Sept., 408.
13. Landolt, D., Muller, R. H. and Tobias, C. W., *J. Electrochem. Soc.* (1969) **116**, No. 10, 1384.
14. Hoar, T. P., *Corrosion Sci.* (1967) **1**, 341.
15. Reddy, A. K. N., Rao, M. G. B. and Bockris, J. O'M., *J. Phys. Chem.* (1965) **42**, 2246.
16. Freer, H. E., Hanley, J. B. and MacLellan, G. D. S., in *Fundamentals of Electrochemical Machining* (see Ref. 8), p. 103.
17. Powers, R. W. and Wilfore, J. F., in *Fundamentals of Electrochemical Machining* (see Ref. 8), p. 135.
18. McGeough, J. A., *Int. J. Prod. Res.* (1971) **9**, No. 2, 311.
19. LaBoda, M. A. and McMillan, M. L., *Electrochem. Technol.* (1967) **5**, 340, 346.
20. Postlethwaite, J. and Kell, A., *J. Electrochem. Soc.* (1972) **119**, No. 10, 1351.
21. Cooper, J., Muller, R. H. and Tobias, C. W., in *Fundamentals of Electrochemical Machining* (see Ref. 8), p. 300.
22. Muller, R. H., Paper presented at First Int. Conf. on ECM, Leicester Univ., March 1973.
23. Brook, P. A. and Iqpal, Q., *J. Electrochem. Soc.* (1969) **116**, No. 10, 1458.
24. Chin, D. T. and Wallace, A. J., *J. Electrochem. Soc.* (1971) **118**, No. 5, 831.
25. Landolt, D., *J. Electrochem. Soc.* (1972) **119**, No. 6, 708.
26. Hoare, J. P., LaBoda, M. A., McMillan, M. L. and Wallace, A. J., *J. Electrochem. Soc.* (1969) **116**, No. 2, 199.

27. Hoare, J. P., *Nature* (1968) **219**, Sept. 7, 1034.
28. Chin, D. T., in *Fundamentals of Electrochemical Machining* (see Ref. 8), p. 250.
29. Mao, K. W., Hoare, J. P. and Wallace, A. J., *J. Electrochem. Soc.* (1972) **119**, No. 4, 419.
30. Chin, D. T., *J. Electrochem. Soc.* (1972) **119**, No. 8, 1043.
31. Boden, P. J. and Evans, J. M., *Nature* (1969) 222, 337.
32. Boden, P. J. and Evans, J. M., *Electrochim. Acta* (1971) **16**, 1071.
33. LaBoda, M. A., Chartrand, A. J., Hoare, J. P., Wiese, C. R. and Mao, K. W., *J. Electrochem. Soc.* (1973) **120**, No. 5, 646.
34. Evans, J. M. and Boden, P. J. in *Fundamentals of Electrochemical Machining* (see Ref. 8), p. 40.
35. Fitz-Gerald, J. M. and McGeough, J. A., Paper presented at I.E.E. Conf. on Electrical Methods of Machining, Forming and Coating, London. Conf. Public. No. 61 (1970) p. 72.
36. Hopenfeld, J. and Cole, R. R., *Trans. A.S.M.E. Series B, J. Eng. Ind.* (1969) Aug., 755.
37. De la Rue, R. E. and Tobias, C. W., *J. Electrochem. Soc.* (1959) Sept., 827.
38. Wagner, C., in *Advances in Electrochemistry and Electro-Chemical Engineering* (ed. P. Delahay and C. W. Tobias), Interscience (1962) **2**, 1.
39. Cook, N. H., Loutrel, S. P. and Meslink, M. C., *Increasing ECM Rates,* Mass. Inst. of Tech. Report (1967).
40. Lovering, D. G., Paper presented at 23rd Meeting, I.S.E., Stockholm, Aug.-Sept. 1972.
41. Gurklis, J. A., *Metal Removal by Electrochemical Methods and Its Effect on Mechanical Properties of Metals*, Defence Metals Information Center, Report 213, Battelle Memorial Institute (1965) Jan.
42. Evans, J. M., Boden, P. J. and Baker, A. A., Proc. 12th Int. Mach. Tool Des. Res. Conf., Macmillan (1971) p. 271.
43. Zimmel, L. J., Paper presented at U.S. Nat. Aero- and Space Eng. Meeting, Oct. 1964.
44. Bannard, J., Paper presented at First Int. Conf. on ECM, Leicester Univ., Mar. 1973.
45. Bayer, J., Cummings, M. A. and Jollis, A. U., Gen. Electric Co., Final Report on Electrolytic Machining Development, Rep. No. ML-TDR-64-313 (1964) Sept.

CHAPTER FIVE

Dynamics and Kinematics

Studies of the behaviour of metal–electrolyte combinations for the conditions of ECM must eventually be utilised in applications of the process. In these applications, certain dynamical and kinematic features of the process are invariably exhibited. If one electrode, the cathode, say, is fed towards the other, the width of the gap between the electrodes will eventually become steady.* This equilibrium gap is a consequence of an inherent feature of ECM whereby the rate of forward movement of the cathode becomes matched by the local

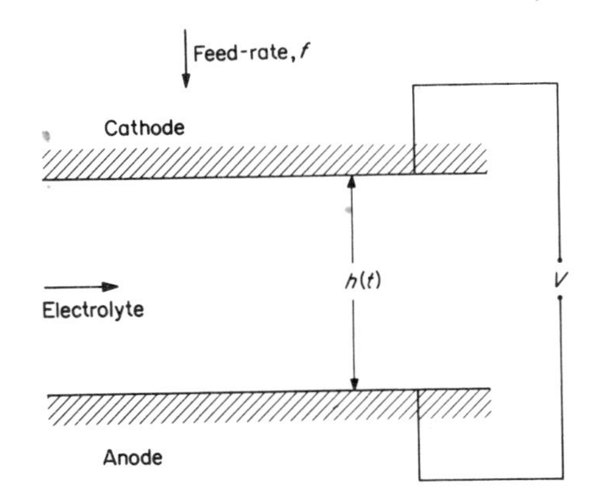

Fig. 5.1 Set of plane parallel electrodes

* The inter-electrode gap referred to here and elsewhere in the text is occasionally referred to as the 'end gap'. Cases in which some other gap is considered should be clear from the discussion at the appropriate part of the text.

anodic dissolution rate. If the cathode is not so moved, the gap will increase indefinitely as machining proceeds.

But these trends of behaviour are subject to other influences. Cathodic gas generation and electrical heating, due to the passage of current, can cause the width of the equilibrium gap to become tapered along the length of the electrode. These two processes have diverse other effects in ECM. Moreover, if they are not controlled, they can lead to the premature termination of machining, although this condition can also be produced by other electrochemical and hydrodynamic factors.

5.1 Variation of gap width with machining time

5.1.1 Direct current

Consider a set of plane parallel electrodes with a constant potential difference V applied across them and with the cathode moving towards the anode at a constant rate f. The electrode configuration is shown in Fig. 5.1. The electrolyte is assumed to be flowing at such a rate that its conductivity in the gap remains constant. For the present, the effect of overpotentials will not be considered, and field effects at the edges of the electrode are ignored. Under these conditions, the width of the gap between the electrodes can be taken to be dependent only on machining time. Let the gap at the start of machining be $h(0)$, and after time t let it be $h(t)$. If 100% current efficiency is also assumed, then from Faraday's law, the rate of change of gap relative to the cathode surface is

$$\frac{\mathrm{d}h}{\mathrm{d}t} = \frac{AJ}{z\rho_a F} - f \tag{5.1}$$

where, as defined previously, A, z, and ρ_a are the atomic weight, valency, and density of the anode metal, J is the current density, and F is Faraday's constant.

For a constant electrolyte conductivity, Ohm's law gives

$$J = \frac{\kappa_e V}{h} \tag{5.2}$$

On substitution into Equation (5.1):

$$\frac{\mathrm{d}h}{\mathrm{d}t} = \frac{MV}{h} - f \tag{5.3}$$

where M $(= A\kappa_e/zF\rho_a)$ is a machining parameter.

Equation (5.3) has the solution

$$t = \frac{1}{f}\left[h(0) - h(t) + h_e \ln \frac{h(0) - h_e}{h(t) - h_e}\right] \tag{5.4}$$

where

$$h_e = \frac{MV}{f} \tag{5.5}$$

is the equilibrium gap, the steady-state solution to Equation (5.3) for $dh/dt = 0$. A useful relationship for determining equilibrium feed-rates can also be obtained from Equations (5.2) and (5.5):

$$f = \frac{e_a J}{\rho_a} \tag{5.6}$$

where e_a $(= A/zF)$ is the electrochemical equivalent for the anode metal.

Three practical cases are now of interest:

(a) When $f = 0$, that is, no cathode movement, Equation (5.3) has the solution

$$h^2(t) = h^2(0) + 2MVt \tag{5.7}$$

The gap then increases indefinitely with the square root of the machining time (Fig. 5.2).

(b) An ever increasing gap is not usually desirable in ECM, particularly when an accurate shape has to be produced on the anode. Accordingly, fixed mechanical movement of the cathode is much more common. Thus, when Equation (5.4) is used to compare the gap

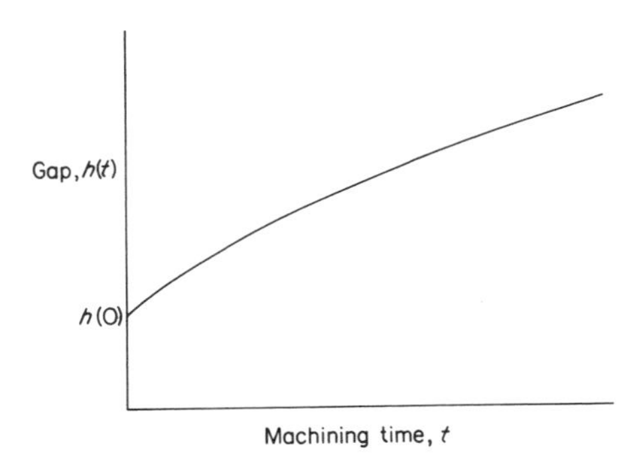

Fig. 5.2 Indefinite increase of gap with machining time (zero feed-rate)

width with machining time, and if the initial gap is greater than the final gap, the gap width is seen to decrease with time to its equilibrium value. When the gap is initially smaller than the steady-state value, it

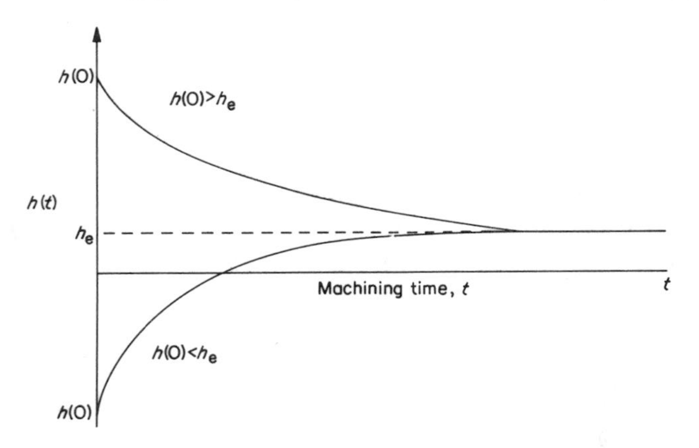

Fig. 5.3 Variation of gap width with machining time (constant feed-rate)

increases to the equilibrium width. This dependence of gap width on machining time is illustrated in Fig. 5.3.
(c) Finally, the condition $h(t) = 0$ implies a short circuit between the electrodes; and values of $h(t)$ less than zero are physically impossible.

Example

For typical values, $e_a = 29 \times 10^{-5}$ g/C, $\kappa_e = 0{\cdot}2$ ohm^{-1} cm^{-1}, $\rho_a = 8$ g/cm^3, $V = 10$ V, and $f = 1{\cdot}66 \times 10^{-2}$ mm/s, we obtain from Equation (5.5) that $h_e = 0{\cdot}44$ mm.

In Equation (5.6), suppose that $J = 50$ A/cm^2, with the same values for e_a and ρ_a as above. Then f is calculated to be $1{\cdot}8 \times 10^{-2}$ mm/s.

The above analyses have been carried out without the complications of overpotentials at the electrodes. In Chapter 3 it was proposed that for ECM conditions the concentration overpotential may be considered to be negligible and that a Tafel expression may be used for the activation overpotential. In the Tafel equation, this overpotential varies only slowly with current density, and in most ECM analyses it is considered constant. Equation (5.5) may then be modified to account for overpotentials:

$$h_e = \frac{M(V - \Delta V)}{f} \tag{5.8}$$

where ΔV describes the sum of the electrode overpotentials, including reversible potentials, and it is assumed to have a constant value.

Example

Suppose that the sum of the overpotentials and reversible potentials comes to 2 V. From Equation (5.8), the corresponding value for h_e is found to be 0·35 mm, the same values being retained for the other process variables used in the previous example.

Finally, a non-dimensionalised treatment of these basic equations is of interest here, in which a non-dimensional function ψ has been defined:

$$\psi = \frac{fh(0)}{\left[\dfrac{A\kappa_e(V - \Delta V)}{zF\rho_a}\right]} \tag{5.9}$$

This function includes all the dynamic and electrochemical features of the process for the conditions stated above. For the cases $\psi < 1$ and $\psi > 1$, the gap width respectively increases and decreases with machining time to the equilibrium gap width, and for $\psi = 1$, the gap width always has its equilibrium value [1].

5.1.2 *A practical case*

The practice of electrochemical deep-hole drilling is a case where part of an anode surface can be machined under conditions of a

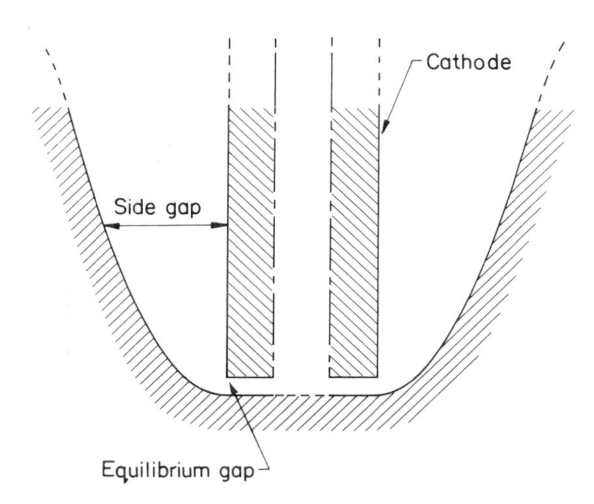

Fig. 5.4 Electrochemical drilling; forward gap (equilibrium) and side gap

constant cathode feed-rate, whilst another part is subject to conditions of zero effective feed-rate. Figure 5.4 illustrates such a case for a cylindrical cathode moving into an anode, the direction of the cathode feed-rate being along the axis of the cathode.

Suppose, first, that machining only takes place over that region of the anode directly adjacent to the leading edge of the cathode. Since ECM there is in the direction of cathode feed, an equilibrium gap will be obtained under the conditions specified by Equation (5.5).

Next, consider dissolution only in the radial direction from the cathode, with ECM in the region of its leading edge ignored. The effective feed-rate normal to the cathode surface is zero, and under that condition, dissolution of the side walls of the anode takes place. The local difference between (i) the radial length between the side wall of the anode and the central axis of the cathode, and (ii) the external radius of the cathode is known as the overcut. The amount of overcut can be diminished by several devices, including insulation along the external side walls of the cathode, so that closer accuracy of machining can be maintained. The relationship between the overcut and the equilibium gap at the leading edge, for both insulated and non-insulated cathode side walls, is discussed in Chapter 7. (It should be remembered that other effects, e.g. those due to electrolyte flow and gas generation, have been omitted in this brief presentation of drilling.)

5.1.3 Alternating current

Although direct current (d.c.) is, by far, more common in ECM, its replacement by alternating current (a.c.) has received some attention [2–4]. The apparent advantage of the use of the latter type of supply is that the need for rectification of the very high currents is abolished. However, on economic grounds, the practical use of a.c. appears to require two further conditions. First, cathode materials such as titanium diboride have to be used, which rectify, at least in part, the current in the electrolytic cell. Secondly, some form of twin cell machining is essential so that alternate dissolution of two anodes can take place.

In this section, the discussion of alternating current is restricted to a simple analysis of the dependence of gap width on machining time. With reference to Fig. 5.1, the constant potential difference V is now assumed to be replaced by an applied voltage, of sinusoidal waveform and of r.m.s. value V_{rms}. The cathode material is considered to be inert; it is not dissolved during the reverse part of the current cycle for which it is anodic. The other assumptions made above are retained with one main exception. Since machining can only take place during one half of each time cycle, the maximum current efficiency can only approach 50%. Experiments have shown that the efficiency is often considerably less, and values about 20 to 30% are common.

For the portion of the time cycle during which the true anode is machined, the rate of change of gap width relative to the cathode surface is given, from Faraday's law, as

$$\frac{\mathrm{d}h}{\mathrm{d}t} = \frac{e_a J \Gamma}{\rho_a} - f \qquad (5.10)$$

where Γ is the current efficiency.

The effective applied voltage V can be expressed in terms of V_{rms}:

$$V = (\sqrt{2}\, V_{rms} - \Delta V) \sin \omega t \qquad (5.11)$$

where ΔV is an expression, assumed constant, for the sum of the reversible potentials and overpotentials at the electrodes.

By use of Ohm's law, Equation (5.10) becomes

$$\frac{\mathrm{d}h}{\mathrm{d}t} = \frac{(\sqrt{2}\, V_{rms} - \Delta V) e_a \kappa_e \Gamma \sin \omega t}{h \rho_a} - f \qquad (5.12)$$

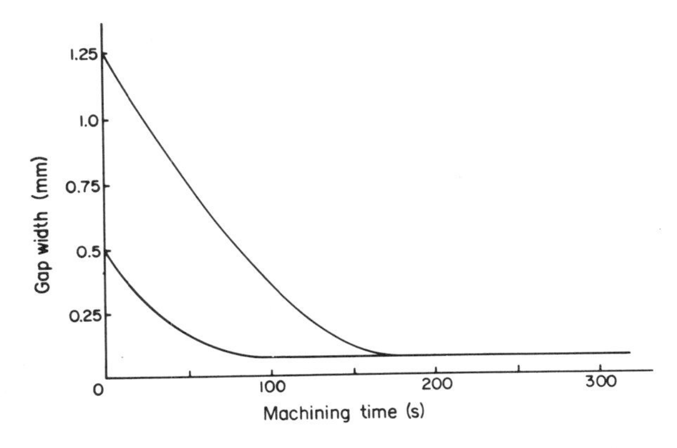

Fig. 5.5 Variation of gap width with machining time (alternating current); $e_a = 25 \times 10^{-5}$ g/C; $\Gamma = 0{\cdot}26$; $\rho_a = 7{\cdot}7$ g/cm^3; $\kappa_e = 0{\cdot}195$ ohm^{-1} cm^{-1}; $V_{rms} = 12$ V; $\Delta V = 1{\cdot}4$ V; $f = 10^{-2}$ mm/s; $h(0) = 1{\cdot}25, 0{\cdot}5$ mm

During the other part of the time cycle for which the true cathode becomes the positive electrode,

$$\frac{\mathrm{d}h}{\mathrm{d}t} = -f \tag{5.13}$$

These equations can be solved numerically to give the gap as a function of time. The results for typical values given in Fig. 5.5 demonstrate that a steady-state gap is achieved in a.c. ECM. The existence of equilibrium gap conditions in a.c. ECM has been confirmed experimentally [2], the above theory being shown to provide a reasonable description of a.c. ECM, despite the number of simplifying assumptions.

5.2 Hydrogen evolution

So far, the influence of the other processes at work in ECM has not been considered. For example, the usual complementary reaction at the cathode, hydrogen evolution, usually reduces the effective conductivity of the electrolyte so that the local anodic dissolution rate varies downstream until equilibrium is achieved. We have seen earlier that the passage of current increases the conductivity by Joule heating, so lessening the effect due to hydrogen. The effects of

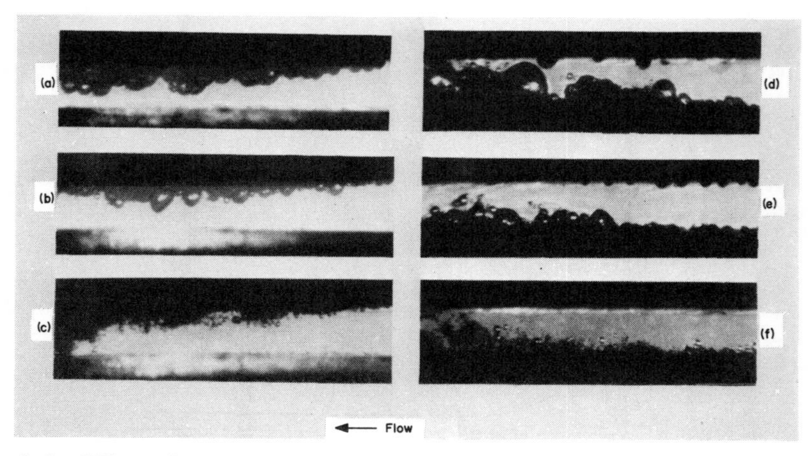

Fig. 5.6 Effect of electrolyte flow velocity and cathode orientation on gas generation in 2N KCl; current density 50 A/cm², electrode gap 0·5 mm. Cathode face down: flow velocity (a) 1 m/s, (b) 2 m/s, (c) 4 m/s. Cathode face up: flow velocity (d) 1 m/s, (e) 2 m/s, (f) 4 m/s. (By permission of Landolt, Acosta, Muller and Tobias [5])

these processes can be analysed from a study of the transport processes appropriate to ECM. It should help in these studies to have some appreciation of (i) the location of bubbles in the machining gap, and (ii) the effects on the size and location of the bubbles of process variables. To this end, some photographic investigations are of interest [5–9].

The formation next to the cathode surface of the hydrogen gas bubble layer can be seen from photographs of the electrode gap region (Fig. 5.6); the related experimental data are also given. The thickness of the bubble layer increases in the downstream direction along the gap, and it decreases as the electrolyte flow-rate is increased. Further photographs demonstrate that the size of individual bubbles also decreases rapidly as the electrolyte velocity is increased from

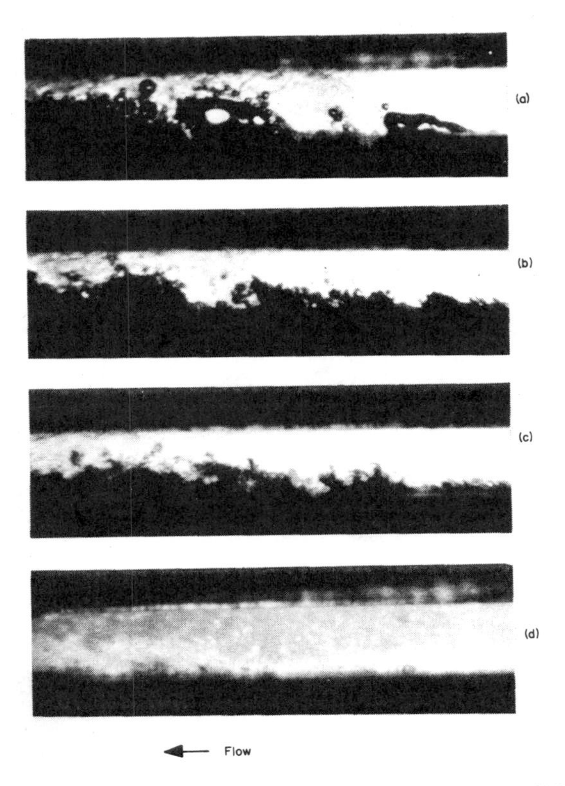

Fig. 5.7 Effect of flow velocity on hydrogen gas evolution in 2N KCl; current density 100 A/cm^2. Cathode face up: flow velocity (a) 4 m/s, (b) 6 m/s, (c) 10 m/s, (d) 25 m/s. (By permission of Landolt, Acosta, Muller, and Tobias [5])

4 to 25 m/s (current density, 100 A/cm^2) (Fig. 5.7). From Fig. 5.8, the bubble size is seen to increase with increasing current density, the flow-rate being held constant.

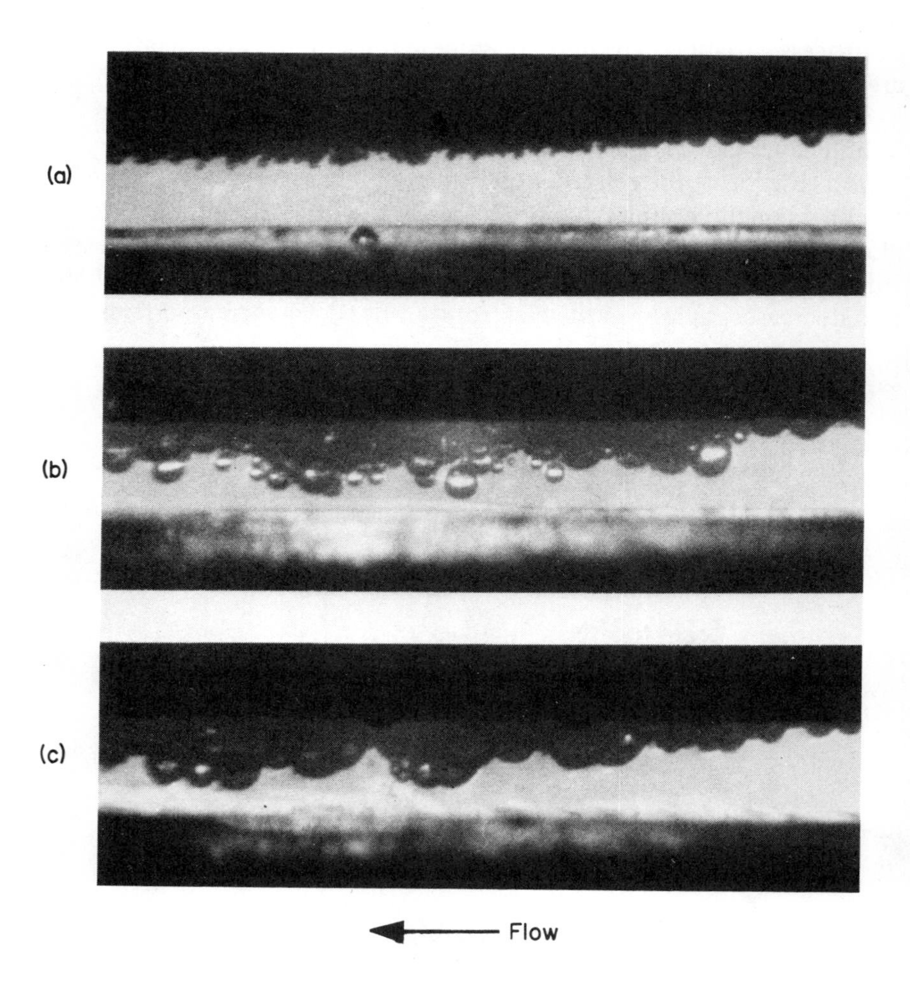

Fig. 5.8 Effect of current density on cathodic gas generation in 2N KCl. Cathode face down. Electrolyte velocity 1 m/s; current density (a) 5 A/cm^2, (b) 20 A/cm^2, (c) 50 A/cm^2 (By permission of Landolt, Acosta, Muller, and Tobias [5])

These observations show that the gas bubbles occupy a region adjacent to the cathode, and that they are not dispersed uniformly throughout the machining gap. From Fig. 5.6, the thickness of the

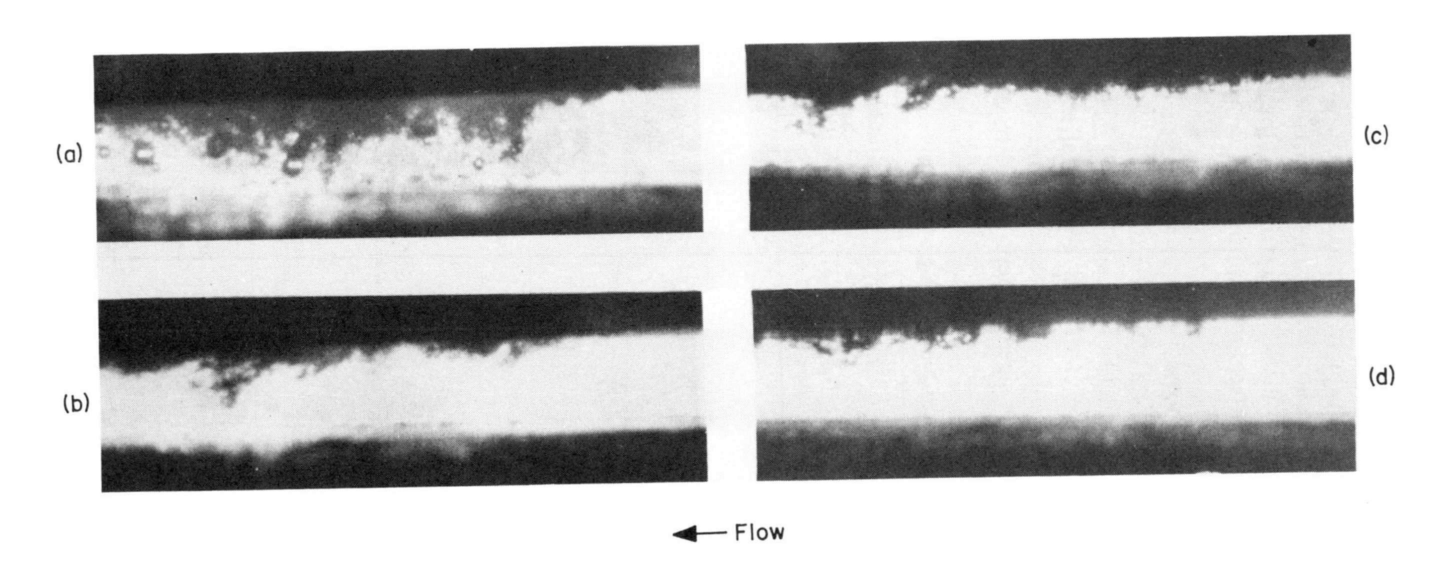

Fig. 5.9 Cathodic gas generation in 2N KNO_3 and 2N KCl. Current density 100 A/cm^2. Cathode face down. Electrolyte velocity (KCl) (a) 4 m/s, (b) 8 m/s; (KNO_3), (c) 4 m/s, (d) 8 m/s (By permission of Landolt, Acosta, Muller, and Tobias [5])

two-phase region is noted to be independent of orientation of the cathode for high velocities (above 1 m/s).

In Chapter 4, some attention was paid to reports that very little cathodic gas is produced in nitrate electrolytes. Observational proof is given in Fig. 5.9, for identical machining conditions in chloride and nitrate electrolytes.

In the work described above, the absolute pressure at the electrodes was about atmospheric and its effect on bubble size was negligible [5]. The relationship between bubble size, mean flow velocity, and absolute pressure has also been studied, for velocities and pressures ranging from about 4 to 18 m/s and 2 to 30 atm respectively [8, 9]. From this study, the bubble size has been found to decrease also

Table 5.1 Mean bubble diameter as a function of process variables

Current density (A/cm^2)	Electrolyte velocity (m/s)	Absolute pressure (approx) (atm)	Temperature (°C)	Electrolyte	Bubble diameter (approx) (μm)	Other information	Ref.
	4·7	32	27	NaCl	50	Mild steel anode, brass cathode, no hydrodynamic entry length	9
		16			62		
		8			81		
		6			90		
		2·5			130		
	10·8	32	27	NaCl	35		9
		16			45		
		8			55		
		6			62		
		2·5			75		
	18	16			27		9
		6			33		
		4			40		
		2·5			47		
50	1	1–1·3	25	2N KCl	99	copper electrodes	5
50	2	1–1·3	25	2N KCl	69		
50	4	1–1·3	25	2N KCl	35		
5	1	1–1·3	25	2N KCl	56	copper electrodes	5
20	1	1–1·3	25	2N KCl	78		
50	1	1–1·3	25	2N KCl	99		

with increasing pressure and velocity; for any given velocity, the bubble diameter appears to be proportional to $(\text{pressure})^{-0\cdot 3}$

Table 5.1 carries a summary of experimental data on bubble size for ECM conditions. The table again confirms that bubble size increases with increasing current density and decreases with increasing electrolyte velocity and absolute pressure.

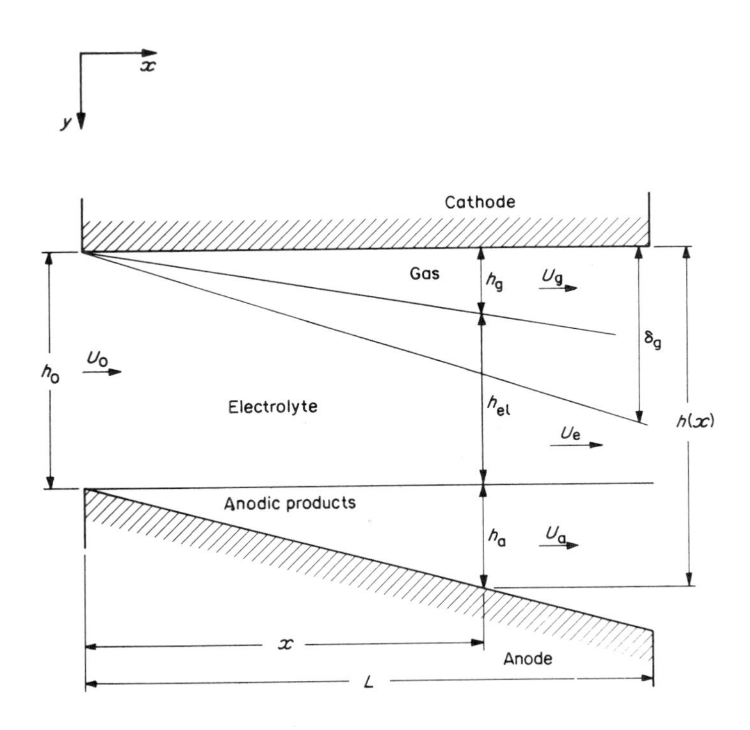

Fig. 5.10 Variation in gap width (after Thorpe and Zerkle [10])

Although the discussion now is concerned with gas evolved at the cathode, it is relevant to recollect here that gas can be evolved at the anode. Current efficiency calculations for gas generation at both electrodes have been discussed earlier in Chapter 4. It should be recalled that bubbles may also appear in the gap due to cavitation. This phenomenon was discussed earlier in Chapter 2. In subsequent sections of the present chapter, only hydrogen gas evolution at the cathode is considered, and 100% efficiency is assumed.

5.3 Transport equations

The principal processes at work in ECM were discussed in Chapters 1 to 3, and many aspects of their bearing on the rate and manner of the electrode reactions were brought to light in Chapter 4.

There are, of course, many other effects, and some are now studied which influence the dynamic and kinematic behaviour of the process. In particular, we shall examine some effects due to hydrogen gas evolution and electrical heating. Initially, the discussion will be restricted to modifications which they impose on the shape of the equilibrium gap. That the effective conductivity of the electrolyte solution is changed sufficiently by gas and heating to produce a tapering of an otherwise parallel gap is often a major problem in cathode design.

This problem can be tackled with the aid of the appropriate transport equations of mass, energy, and electrical charge in ECM, several simplified models of which are available [7, 10–12]. Of these models, one yields essentially a computer solution of the transport equations [7], whilst the others supply an analytic solution. We follow the latter argument, keeping in mind that the model incorporates some assumptions which may not necessarily apply in practice and that it omits some physical features of ECM. Nevertheless, much useful information can still be deduced from it.

5.3.1 Introductory equations

Figure 5.10 accounts for the modified shape of the equilibrium gap. Since steady-state machining conditions are assumed, the gap does not change further with machining time. But its equilibrium width does now vary with distance x along the electrode length. As only equilibrium conditions are considered, we can also drop the subscript e without loss of meaning; the gap now varies from its width at inlet, h_0, in some undefined fashion to its value $h(x)$ downstream. The cathode still moves towards the anode at a fixed rate f. The electrolyte velocity is assumed to have values U_0 at inlet and $U(x)$ downstream; the velocity is also assumed to have no variation in the y direction.* The current density J causes hydrogen gas to be evolved in the vicinity of the cathode at a local flux rate m_g (mass

* Since the electrolyte velocity is assumed to have no variation across the electrode gap width, the symbol U is now used for the velocity, in common with the usual notation for ECM.

per unit time per unit area). The hydrogen gas is considered to flow at some rate which is equivalent to the local layer width h_g, the gas bubbles being contained within an electrolyte layer of thickness δ_g. If the bubbles occupy the whole channel, then $\delta_g = h$. This condition may arise at some distance downstream. If the bubbles become completely coalesced, then $\delta_g = h_g$, and they would then be completely separated from the pure electrolyte. (ECM would also be terminated because the cathode would be insulated from the anode.) The anodic reactions result in an equivalent layer (for example, of metal hydroxide) at the anode of local width h_a, the local flux rate being m_a. The central region, of width h_{el} is occupied by pure electrolyte.* The three phases, gas, anodic products, and electrolyte, are assumed to flow in the downstream direction with a local average velocity in some area proportional to the width h. That is,

$$\begin{aligned} U_g &\propto h_g \\ U_e &\propto h_{el} \\ U_a &\propto h_a \end{aligned} \tag{5.14}$$

Also,

$$h(x) = h_g + h_{el} + h_a \tag{5.15}$$

The gas phase fraction, α, and anodic products fraction, β, can now be conveniently defined as

$$\begin{aligned} \alpha &= \frac{h_g}{h} \\ \beta &= \frac{h_a}{h} \end{aligned} \tag{5.16}$$

The equivalent relationship for the electrolyte then is

$$(1 - \alpha - \beta) = \frac{h_{el}}{h} \tag{5.17}$$

* To avoid confusion with the recognised symbol for the equilibrium gap, h_e, the subscript 'el' is used here to identify the thickness of the layer of pure electrolyte. The subscript 'e' is retained to denote the other properties of the pure electrolyte layer.

Corresponding fractions for the mass flow-rates of hydrogen gas and anodic products can be expressed in terms of the individual mass flow-rates:

$$\mu = \frac{\omega_g}{\omega}$$

$$\nu = \frac{\omega_a}{\omega} \tag{5.18}$$

$$(1 - \mu - \nu) = \frac{\omega_e}{\omega}$$

where

$$\omega = \omega_g + \omega_e + \omega_a \tag{5.19}$$

$$\omega_g = \rho_g h_g U_g$$

Similar expressions apply for ω_a and ω_e.

The slip ratios σ and Ω are defined by

$$\sigma = \frac{U_g}{U_e}$$

$$\Omega = \frac{U_a}{U_e} \tag{5.20}$$

Certain approximations, which are valid in practice, can ease the subsequent analysis. In Chapter 4, it was stated that the anodic products in solution have negligible effect on the density and conductivity of the bulk solution. Moreover, Thorpe and Zerkle have proposed that, for inlet electrolyte pressures of the order of 690 kN/m^2 (which are typical for ECM), β is much smaller than α. It is reasonable then to assume that h_a, U_a, and β are negligible.

Before we proceed further, it may be helpful to identify more fully the notion of the mass flux rate m_a. From Faraday's law,

$$m_a = e_a J \tag{5.21}$$

Since, for equilibrium ECM,

$$J = \frac{\rho_a f}{e_a} \tag{5.6}$$

we have

$$m_a = \rho_a f \tag{5.22}$$

which is a measure of the local rate of recession of the anode surface in the direction normal to the cathode surface.

5.3.2 Equation of continuity of mass

A volume control element in the inter-electrode gap is shown in Fig. 5.11. Since the anodic products are considered to have negligible effect, only the hydrogen gas and pure electrolyte phases are shown.

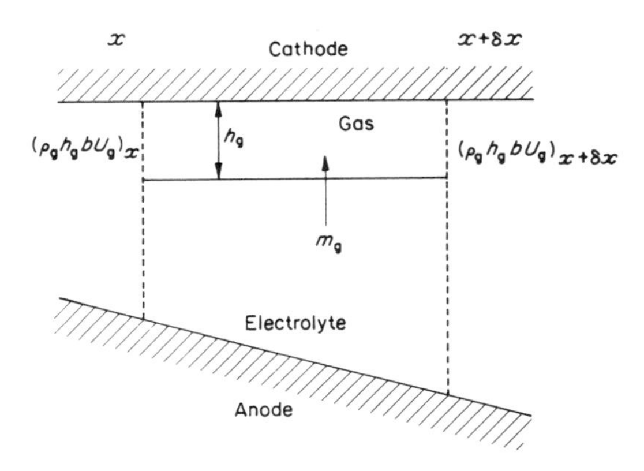

Fig. 5.11 Control volume element for continuity of mass

They are assumed to occupy the regions indicated. For the purposes of the analysis, the hydrogen gas is assumed to be evolved at the interface between the electrolyte and gas phases, and to enter the gas control element. If the principle of conservation of mass for flow in a duct is applied, analysis yields the equation of continuity of mass flow from section to section.

Applying the law of conservation of mass to the gas-phase control volume, we see that

$$(\rho_g h_g b U_g)_x + m_g \delta x b = (\rho_g h_g b U_g)_{x+\delta x} \tag{5.23}$$

where b is the electrode width. In the limit $\delta x \to 0$, this becomes

$$\frac{d}{dx}(\rho_g h_g U_g) = m_g \tag{5.24}$$

From Equations (5.16) and (5.20), and since

$$m_g = e_g J$$

we obtain

$$\frac{d}{dx}(\rho_g \alpha \sigma h U_e) = e_g J \tag{5.25}$$

A similar equation can be derived for the electrolyte component:

$$\frac{d}{dx}[\rho_e(1 - \alpha)h U_e] = (e_a - e_g)J \tag{5.26}$$

5.3.3 Energy equation

Consider the volume element for energy balance shown in Fig. 5.12. Hopenfeld and Cole [7] have enumerated the relevant factors which contribute to generation of heat within the volume element: (a) Joule heating, (b) irreversible energy released from electrochemical reactions at the surfaces of the electrodes, (c) energy due to secondary chemical reactions in the bulk of the electrolyte, (d) the reversible heat of the electrochemical reactions, (e) heat lost or added to the fluid from the surrounding surfaces, e.g. the walls of the electrodes, and (f) the mechanical heat generated in the fluid by the viscous component of the flow. Any subsequent analysis can be considerably simplified if the reasonable assumption is made that the heat arising from the contributions (b) to (f) is negligible compared with that from (a). If the heat capacity and thermal conductivity of the gas phase are also assumed to be sufficiently low, the hydrogen can be assumed to absorb none of the electrical energy

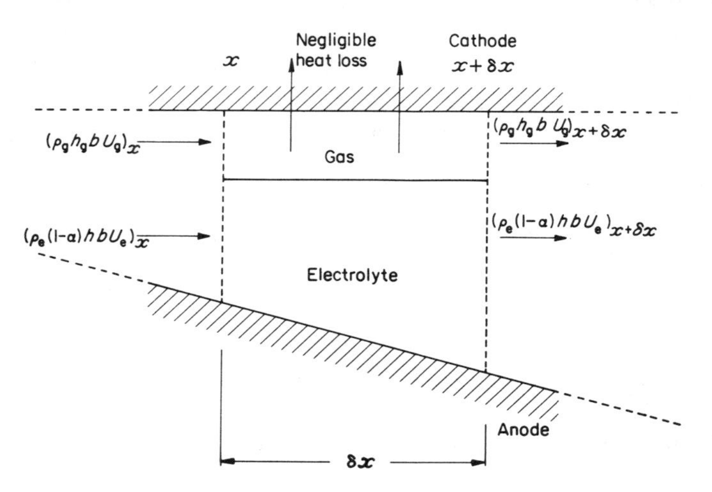

Fig. 5.12 Control volume element for continuity of energy

dissipated between the electrodes. Then only the liquid electrolyte phase has to be considered.

Assuming now that the electrical energy added to the volume element is dissipated as heat in the electrolyte, we deduce the temperature gradient,

$$\frac{\mathrm{d}T}{\mathrm{d}x} = \frac{J(V - \Delta V)}{c_e(1 - \alpha)h\rho_e U_e} \tag{5.27}$$

where c_e is the specific heat of the electrolyte.

Thorpe and Zerkle also give the equations of state for the gas and electrolyte as

$$\rho_g = \frac{p}{RT} \tag{5.28}$$

$$\rho_e = \rho_0 \tag{5.29}$$

where p is the static pressure, R is the Gas Constant, and T is the temperature. The suffices g and e denote the quantity for the gas and electrolyte whilst the suffix 0 denotes inlet conditions.

5.3.4 *Equation for balance of electrical charge*

For the electrical charge across the volume element to be balanced, its distribution at the metal–electrolyte interfaces, across the diffusion layers, and at the outer boundaries of the diffusion layers must be known, as well as the particular electrochemical reaction. This information is difficult to find. Indeed, Bass [13] has pointed out the difficulties of balancing the charge just within the diffusion layer with the charge in the region just outside it. A commonly accepted, simpler approach is to use a form of Ohm's law to describe the balance of charge across the gap:

$$J = \frac{\kappa_m(V - \Delta V)}{h} \tag{5.30}$$

where κ_m is the conductivity of the electrolyte–gas medium. Next, as proposed by Thorpe and Zerkle, the gas-phase fraction is regarded as equivalent to the void fraction. Then, from the work of De la Rue and Tobias [14], the conductivity κ_m can be expressed further in terms of the electrolyte conductivity, κ_e, and the void fraction, α, by use of the Bruggeman equation:

$$\kappa_m = \kappa_e(1 - \alpha)^n \tag{5.31}$$

The power n is normally 1·5. For α sufficiently small, say $\alpha < 0{\cdot}2$, this equation can be approximated by

$$\kappa_m = \kappa_e(1 - 1{\cdot}5\alpha) \tag{5.32}$$

An expression can also be obtained from Ohm's law which relates κ_m and κ_e with the thickness δ_g of the layer containing the hydrogen bubbles:

$$(V - \Delta V) = \frac{J\delta_g}{\kappa_m} + \frac{J(h - \delta_g)}{\kappa_e} \tag{5.33}$$

Here it is assumed that $\delta_g = h$ at outlet.

It has been pointed out earlier that the conductivity κ_e also depends on temperature:

$$\kappa_e = \kappa_0[1 + \zeta(T(x) - T_0)] \tag{5.34}$$

where κ_0 is the conductivity of the electrolyte at inlet, ζ is the temperature coefficient for conductivity, $T(x)$ is the temperature at some control point, and T_0 is the temperature at inlet.

5.3.5 *Discussion of quasi-steady processes*

The stage has now almost been reached where the transport equations can be used to investigate the modifications imposed on the shape of the equilibrium gap by the gas and electrical heating. Before this can be done, however, one other matter needs consideration. The void fraction, the solution temperature, and the velocity themselves have a response to the changes in the gap conditions caused by the gas and heating. We must, therefore, examine the time taken for the void fraction, the solution temperature, and velocity within the gap to reach an equilibrium state.

In the study of this problem, it is convenient to write the governing kinematic equation (5.3) in the form

$$\frac{1}{f}\,\frac{\partial h}{\partial t} = \frac{J(t)}{J} - 1$$

where for clarity the symbol $J(t)$ is now a time-dependent current density whilst the symbol J is retained for the equilibrium current density:

$$J = \lim_{t\to\infty} J(t) = \frac{f\rho_a}{e_a}$$

Previously, the void fraction and slip ratio have been defined by

$$\alpha = \frac{h_g}{h} \quad \text{and} \quad \sigma = \frac{U_g}{U_e}$$

Next, a set of characteristic dimensionless variables are introduced which have unit order of magnitude at some location along the gap:

$$\begin{aligned} x^0 &= \frac{x}{L}, \quad t^0 = \frac{ft}{h_0} \\ h^0 &= \frac{h}{h_0}, \quad U^0 = \frac{U}{U_0} \\ T^0 &= \frac{T}{T_0}, \quad \rho^0 = \frac{\rho}{\rho_0}, \quad p^0 = \frac{p}{p_0} \end{aligned} \tag{5.35}$$

Now consider the equation of continuity of mass. It can be readily deduced that its time-dependent form is

$$\frac{\partial}{\partial t}(\rho_g h_g) + \frac{\partial}{\partial x}(\rho_g h_g U_g) = m_g \tag{5.36}$$

Using the above expressions for α and σ,we have

$$\frac{\partial}{\partial t}(\rho_g \alpha h) + \frac{\partial}{\partial x}(\rho_g \alpha h \sigma U_e) = m_g \tag{5.37}$$

By means of the non-dimensional variables, this equation can be written

$$\frac{\partial}{\partial t^0}(\alpha \rho_g^0 h^0) + \frac{h_0 U_0}{Lf} \frac{\partial}{\partial x^0}(\rho_g^0 h^0 U_e^0 \alpha \sigma) = \frac{m_g}{f\rho_0}$$

Defining

$$\gamma = \left(\frac{h_0}{L}\right)\left(\frac{U_0}{f}\right)$$

we obtain

$$\frac{1}{\gamma}\frac{\partial}{\partial t^0}(\alpha \rho_g^0 h^0) + \frac{\partial}{\partial x^0}(\rho_g^0 h^0 U_e^0 \alpha \sigma) = \frac{1}{\gamma}\rho_g^0\left(\frac{J(t)}{J}\right)\left(\frac{e_g}{e_a}\right)\left(\frac{\rho_a}{\rho_g}\right) \tag{5.38}$$

Now the quantities x^0, t^0, h^0, U_e^0, α, σ, $J(t)/J$ are all of unit order of magnitude. That is, all the terms in a non-dimensionalised kinematic equation must be retained.

The non-dimensional quantity γ can be regarded as the ratio of two time constants for the process, h_0/f and L/U_0. The former time constant is a measure of the time required for the cathode to travel a distance equal to the inlet electrode gap; the latter constant is a measure of the time for the electrolyte to travel through the electrode gap. If we insert typical values, we have

$$\frac{h_0}{f} \simeq [0(1)], \frac{L}{U_0} \simeq [0(10^{-3})]$$

i.e.

$$\gamma \simeq [0(10^3)].$$

Thus we conclude from Equation (5.38) that at the feed-rates used in ECM, the time-dependent changes in the mass flow occur instantaneously compared with the changes dependent upon position along the gap. In the same way, the transient terms in the other transport equations can be shown to be negligible.

These results mean that the void fraction, electrolyte temperature, and velocity reach local steady flow conditions almost instantaneously in response to transients associated with changes in the gap width given by the kinematic equation. It should be realised, however, that this analysis does not provide information about the time taken for the current density to reach an equilibrium value along the electrode length.

5.4 Expression for the void fraction

We are now in a position to derive an expression for the void fraction, α.

First, Equations (5.25) and (5.26) are integrated, the boundary conditions at $x = 0$ being,

$$\alpha = 0, \quad h = h_0, \quad \rho_e = \rho_0, \quad U_e = U_0$$

The resulting equations can then be rearranged to give

$$hU_e = \frac{e_g Jx}{\rho_g \sigma \alpha}$$

and

$$hU_e = \frac{(e_a - e_g)Jx + \rho_0 h_0 U_0}{\rho_e (1 - \alpha)}$$

That is,

$$\frac{1-\alpha}{\alpha} = \frac{[(e_a - e_g)Jx + \rho_0 h_0 U_0]\rho_g \sigma}{\rho_e e_g Jx} \tag{5.39}$$

If we use Equation (5.29)

$$\rho_e = \rho_0 \tag{5.29}$$

and, following Thorpe and Zerkle, we define

$$A = \frac{\rho_0 e_g}{\sigma \rho_g e_a} \tag{5.40}$$

then Equation (5.39) becomes

$$\frac{1-\alpha}{\alpha} = \frac{(e_a - e_g)}{Ae_a} + \frac{\rho_0 h_0 U_0}{Ae_a Jx}$$

that is,

$$\alpha = \frac{Ae_a Jx}{\rho_0 h_0 U_0 + Ae_a Jx\left(1 + \dfrac{e_a - e_g}{Ae_a}\right)} \tag{5.41}$$

By introduction of the non-dimensional variables

$$e^* = \frac{e_a - e_g}{e_a} \quad \text{and} \quad x^* = \frac{x}{LS} \tag{5.42}$$

where

$$S = \frac{\rho_0 h_0 U_0}{L\rho_a f} \tag{5.43}$$

and which, on substitution for f from Equation (5.6), becomes

$$S = \frac{\rho_0 h_0 U_0}{Le_a J} \tag{5.44}$$

then Equation (5.41) takes the non-dimensional form

$$\alpha = \frac{Ax^*}{1 + (A + e^*)x^*} \tag{5.45}$$

Note that A depends inversely on the gas density ρ_g whose variation with distance x^* along the electrode has not been established. However, if the pressure drop across the gap is small, A can be assumed to be approximately constant.

5.5 Taper in width of equilibrium gap

5.5.1 Effects due to gas and heating

When Equation (5.26) is integrated, we obtain

$$\rho_e(1-\alpha)hU_e = (e_a - e_g)Jx + \rho_0 h_0 U_0 \tag{5.46}$$

the constant of integration being evaluated from the boundary conditions

$$\rho_e h U_e = \rho_0 h_0 U_0$$

and

$$\alpha = 0$$

at

$$x = 0.$$

If the expression for $(1-\alpha)hU_e$ is substituted from Equation (5.46) into the energy equation (5.27), and if J is also eliminated from Equation (5.27) by use of Equation (5.6), Equation (5.27) becomes

$$\frac{dT}{dx^*} = \left(\frac{V}{e_a c_e}\right)\frac{1}{(1+e^*x^*)} \tag{5.47}$$

Note that for simplicity the constant ΔV has been neglected. The quantity $V/e_a c_e$ can be regarded as a reference temperature, T_r. Equation (5.47) has the solution

$$T = T_0 + \frac{T_r}{e^*}\ln(1+e^*x^*) \tag{5.48}$$

The variation in the equilibrium gap width due to hydrogen gas and heating can now be calculated. This analysis can be simplified, if it is assumed that the hydrogen bubbles completely fill the electrode gap. That is,

$$\delta_g = h$$

If overpotentials are also ignored, Equation (5.33) then reduces to

$$V = \frac{Jh}{\kappa_m}$$

At entry to the gap,

$$V = \frac{Jh_0}{\kappa_0}$$

By expressing κ_m in terms of κ_0 from Equations (5.31) and (5.34), we can then deduce the local gap width:

$$\frac{h}{h_0} = (1 - \alpha)^n [1 + \zeta(T - T_0)] \tag{5.49}$$

From Equations (5.45) and (5.48), the ratio h/h_0 can be expressed in terms of the non-dimensional distance x^*:

$$\frac{h}{h_0} = \left[\frac{1 + e^*x^*}{1 + (A + e^*)x^*}\right]^n \left[1 + \frac{B}{e^*} \ln(1 + e^*x^*)\right] \tag{5.50}$$

where $B = \zeta T_r$

Let us now compare the variations in gap width due to heating and hydrogen gas separately.

5.5.2 Effects due to heating

If the presence of hydrogen gas is ignored, we may put $\alpha = 0$ in Equation (5.49), and substitution for $(T - T_0)$ from Equation (5.48) gives

$$\frac{h}{h_0} = \left[1 + \frac{\zeta T_r}{e^*} \ln(1 + e^*x^*)\right] \tag{5.51}$$

5.5.3 Effects due to hydrogen gas

This variation can be obtained in a similar fashion to that of the above section; ignoring electrolyte heating, we have

$$\frac{h}{h_0} = (1 - a)^n$$

$$= \left[\frac{1 + e^*x^*}{1 + (A + e^*)x^*}\right]^n \tag{5.52}$$

Expressions (5.45), (5.51), and (5.52) can be simplified further by approximations for e^* and e^*x^* which are based on practical values for the process variables.

Consider Equation (5.42):

$$x^* = \frac{\rho_a f x}{\rho_0 U_0 h_0}$$

For the values $\rho_a = 8$ g/cm^3, $\rho_0 \simeq 1$ g/cm^3, $h_0 = 0{\cdot}4$ mm, $U_0 = 1{\cdot}5$ m/s $f = 1{\cdot}66 \times 10^{-2}$ mm/s, e_a 29×10^{-5} g/C, $e_g = 1 \times 10^{-5}$ g/C, $x^* \simeq 2 \times 10^{-3}$ to 2×10^{-2} for x ranging from 10 to 100 mm. Also $e^* < 1$, so that $e^*x^* \ll 1$.

For this condition for e^*x^*, approximations to the solutions (5.45), (5.51), and (5.50) can be deduced:

$$\alpha = \frac{Ax^*}{1 + Ax^*} \tag{5.53}$$

$$\frac{h}{h_0} = (1 + \zeta T_r x^*) \tag{5.54}$$

$$\frac{h}{h_0} = \frac{1 + Bx^*}{(1 + Ax^*)^n} \tag{5.55}$$

From Equation (5.54) note that, with Joule heating only, the gap increases linearly with distance along the electrode. This result is the same as that obtained by Tipton [15] who implicitly used both the energy equation and the approximation that $e^*x^* \ll 1$.

These equations demonstrate the significance of the quantities A and B. From Equation (5.53), A represents the effect of the void fraction, whilst from Equation (5.54), B is a measure of the effect of heating. In Equation (5.55), if $n = 1$ and $A = B$, the effect of the hydrogen gas is balanced by the temperature effect.

5.5.4 Effect due to polarisation

One other effect on the gap width needs consideration. The presence of overpotentials at the electrodes can have the apparent effect of increasing the gap width between the electrodes. This phenomenon is well known in electrodeposition, and in Chapters 6 and 7 an analysis is carried out to establish the characteristic, corresponding behaviour in ECM. Although that analysis demonstrates mainly the effect of overpotentials on the processes of anodic smoothing and shaping, it also shows that the width of the gap is increased by an apparent amount

$$\kappa_e \frac{\partial f(J)}{\partial J}$$

which quantity is evaluated for values of a defined mean current density. Here $f(J)$ is a current density-dependent overpotential at the cathode.

Example

Suppose that the Tafel equation applies at the cathode $\eta_a = a + b \log J$. The apparent increase in gap width due to that overpotential alone is $\kappa_e b/J$. For typical values, $\kappa_e = 0{\cdot}1$ ohm^{-1} cm^{-1}, b (= $2{\cdot}3\ RT/z\alpha F$, where $\alpha \simeq 0{\cdot}5$, $z = 1$, $RT/F = 1/40$ V), $J = 50$ A/cm^2, the apparent increase in gap is about $2{\cdot}3 \times 10^{-3}$ mm. This calculation indicates that the overpotential has little effect in increasing the gap, which, typically, is about 0·4 mm.

5.5.5 ***Further studies of gap variation***

In Fig. 5.13 the theoretical expression (5.55) for the variation in gap width with distance along the electrode length is compared with the experimental results obtained by Hopenfeld and Cole. Both investigations show that the gap becomes convergent in the downstream direction.

Nevertheless, some comments must be made on these results. The theoretical curve was derived on the assumption that the quantity A is constant. This assumption is a reasonable one, provided that the pressure drop along the gap is small. This was the case for the experimental work discussed in relation to Fig. 5.13.

However, at higher pressure drops or lower flow-rates, a constant value for A cannot be assumed, and the simple theory breaks down. (Experimental results for higher pressures are described below.)

The index $n = 1{\cdot}5$ has been used on the basis of a uniform distribution of gas bubbles across the gap. That is, $\delta_g = h$. As has been seen in Section 5.2, however, an accumulation of bubbles close to the cathode surface appears to be more likely. Since a greater reduction of current should then occur, a value of n greater than 1·5 would seem more acceptable. Accordingly, a further curve, for $n = 2$, has also been produced, and, as shown in Fig. 5.13, closer agreement with the experimental results is obtained.

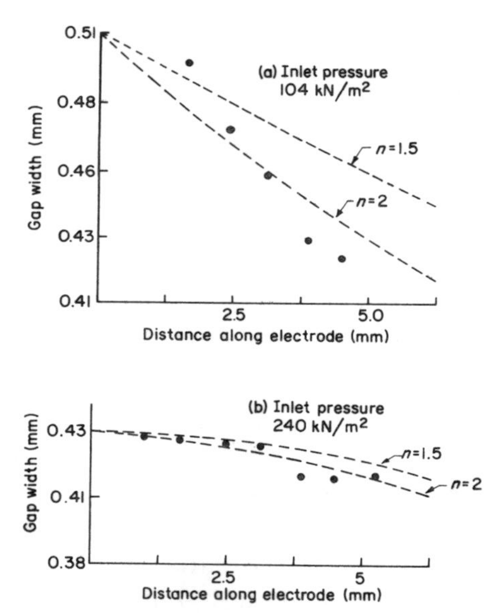

Fig. 5.13 Variation in gap width with distance along the electrode; 0·67N KCl electrolyte; rectangular Al anodes 6·35 mm x 9·53 mm; T_0 = 24°C (a), 27°C (b); V = 19·5 V (a), 19 V (b); $Q_0 = 1{\cdot}4 \times 10^{-8}$ m³/s (a), $6{\cdot}5 \times 10^{-8}$ m³/s (b); $I \simeq$ 19 A (a), 28 A (b) (after Thorpe and Zerkle [10])

Higher pressure drops, from 14 MN/m² at gap inlet to 69 kN/m² at outlet, with correspondingly high electrolyte velocities, have been used in further investigations of the variation in gap shape [16, 17]. The variation in gap found for these conditions is shown in Fig. 5.14. [In that figure both coordinates are plotted in dimensionless form, the suffix 0 denoting the inlet condition. When the gap ratio, $h(x)/h_0 - 1$, is zero there is no taper of the gap, when the ratio is positive the gap is divergent, and when it is negative the gap is convergent.]

The figure shows that the taper is divergent. That is, electrical heating has predominated over hydrogen gas in affecting the effective conductivity of the electrolyte solution, and hence the shape of the gap. The result is attributed to the high electrolyte flow-rates and pressures, which considerably reduce the specific volume of hydrogen gas bubbles. That reduction then permits a greater influence of electrical heating on the effective conductivity of the electrolyte.

These variations in shape of the gap which have been discussed suggest that further investigations of the effects of gas generation and electrical heating are necessary. Much information can stem

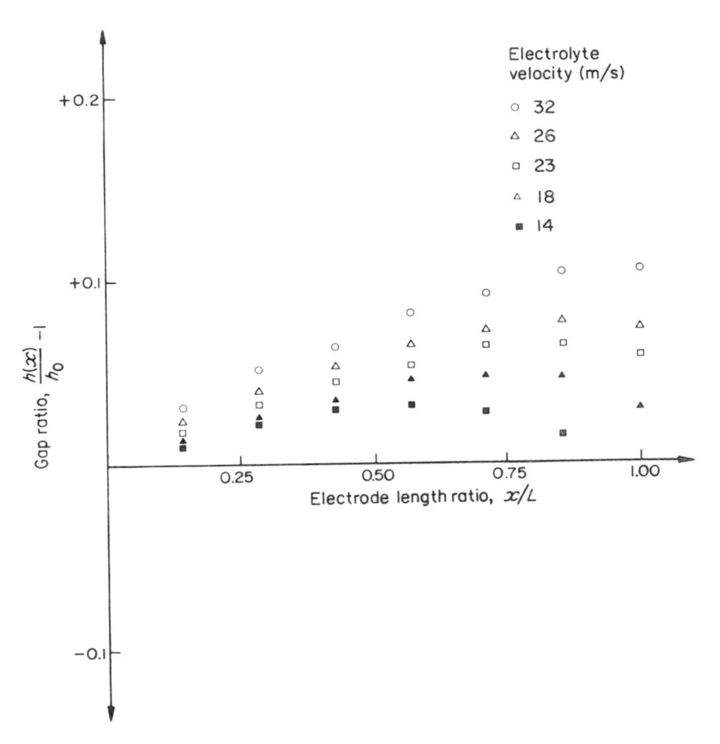

Fig. 5.14 Variation in gap width with electrolyte velocity; 2·5N NaCl; 0·038 m x 0·076 m rectangular electrodes; mild steel anode; $V = 20$ V; $J = 48$ A/cm^2; $f = 16{\cdot}7 \times 10^{-3}$ m/s; $h_0 = 0{\cdot}5$ mm; pressure = 14 MN/m^2 (inlet), 69 KN/m^2 (outlet) (after Clark, McGeough and Yeo [16])

from studies of the velocity, pressure, and temperature distribution of the flowing electrolyte solution during ECM.

5.6 Velocity distribution along electrode length

The evolution of hydrogen and electrical heating also cause a variation in the velocity distribution along the electrode. Thorpe and Zerkle have obtained an analytic expression for this variation, although they do not consider any variation in the regions of the boundary layers. Using Equations (5.26), (5.45), and (5.50), they derive

$$\frac{U_e}{U_0} = \frac{[1 + (e^* + A)x^*]^{n+1}}{(1 + e^*x^*)^n \left[1 + \frac{B}{e^*} \ln(1 + e^*x^*)\right]} \qquad (5.56)$$

The authors point out that if the pressure drop is small compared to ambient pressure, the quantity A, which is really a function of position, will be essentially constant. The equation can be reduced further if, as is usually the case, the coordinate x^* is of order of 10^{-2} or less; since $e^* < 1$, then $e^*x^* \ll 1$. The above equation becomes

$$\frac{U_e}{U_0} = \frac{(1 + Ax^*)^{n+1}}{1 + Bx^*} \qquad (5.57)$$

5.7 Pressure distribution along electrode length

In Chapter 2 expressions were derived for the pressure drop along a rectangular channel for single-phase flow (Sections 2.6 and 2.9). The basis of that work was that established hydrodynamic theory was applicable to (particularly) turbulent flow down small channels of dimensions similar to those met in ECM. In fact, little is known of the hydrodynamic phenomena which may arise under these geometric conditions.

Hopenfeld and Cole[7] have carried out some single-phase flow experiments on the variation of the friction factor with distance along a rectangular channel, of breadth and gap width 9·5 mm and 0·375 mm respectively, for Reynolds numbers ranging from 4 500 to 15 000. Their main results show that, as the Reynolds number is increased, the friction factor decreases, and that for values of the ratio, distance along the electrode to hydraulic diameter, greater than 50, the friction factor coincides with that obtained from the Blasius relation for fully developed turbulent flow. It is emphasised, however, that these experiments were carried out without ECM.

Further information is available in other studies which have indicated that the conventional relationships between Nusselt number and Reynolds number and Schmidt number are applicable to ECM conditions [18]. On that basis, the conventional formulae for pressure drop and diffusion layer thicknesses for linear flow between closely spaced electrodes have been used in the determination of the data in Table 5.2. It is pointed out that these thin layer thicknesses and high rates of transport mean that very high power is required to pump the electrolyte through such small gaps.

Table 5.2 Pressure drop and diffusion layer thickness for linear flow of electrolyte between electrodes of gap 0·5 mm; $\rho_e = 1$ g/cm^3, $\nu = 1$ mm^2/s, $D = 4 \times 10^{-6}$ cm^2/s (after Tobias [18])

Reynolds number	Velocity (m/s)	Pressure drop (atm/cm)	Diffusion layer (mm)
10 000	10	$1{\cdot}5 \times 10^{-1}$	$2{\cdot}0 \times 10^{-3}$
20 000	20	$4{\cdot}9 \times 10^{-1}$	$1{\cdot}1 \times 10^{-3}$
100 000	100	$8{\cdot}2$	$3{\cdot}2 \times 10^{-5}$

In ECM, moreover, the presence in the electrolyte solution of the products of the reactions at both electrodes means that the flow patterns should differ from those of a single-phase flow. Attention has already been drawn in Chapter 2 to one change that arises with these multi-phase flow conditions: a higher inlet pressure is now required to maintain the same rate of flow of electrolyte solution.

5.8 Temperature distribution along electrode length

In the course of the analysis in Section 5.5, an expression was derived for the temperature rise along the electrode length:

$$T(x^*) = T_0 + \frac{T_r}{e^*} \ln(1 + e^* x^*) \tag{5.48}$$

For the usual approximation, $e^*x^* \ll 1$, and on substitution of the process variables, Equation (5.48) reduces to

$$T(x) - T_0 = \frac{VJx}{c_e \rho_e h_0 U_0}$$

where Equation (5.6) has also been used. That is, a linear increase in temperature with distance along the gap should be expected. The assumptions, previously discussed, including low pressure drop along the electrode length, still hold.

The problem of electrical heating of the electrolyte has received much attention, often in relation to the limitations imposed on the process by excessive heating, or boiling. It will be recalled that, in Chapter 1, estimates of the Joule heating of the solution provided a simple means of estimating typical electrolyte velocities in ECM.

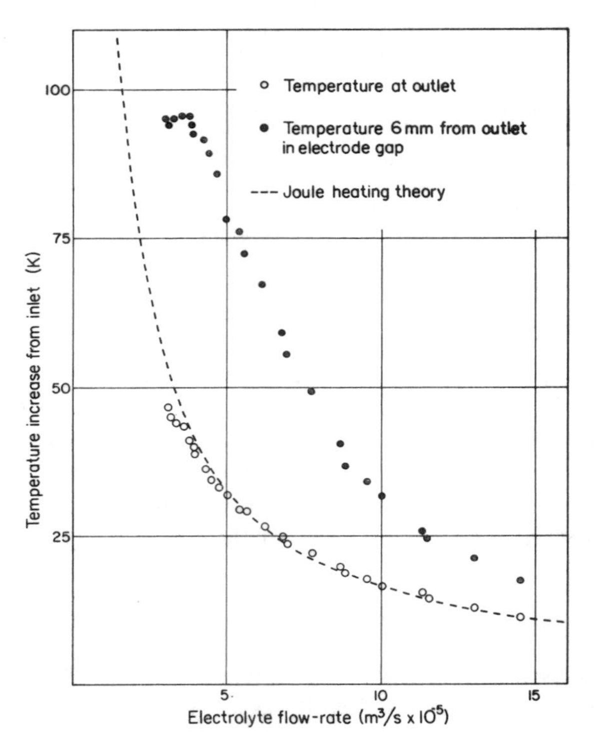

Fig. 5.15 Variation in temperature increase with electrolyte flow-rate; $J = 16$ A/cm^2; $V = 32$ V; pressure = 170 KN/m^2 (outlet); $f = 6 \times 10^{-3}$ mm/s (after Clark and McGeough [19])

Results of measurement of the temperature variation of a 10% (w/w) NaCl solution, flowing down a rectangular channel of length, breadth, and inlet width 100 mm, 13 mm, and 2·4 mm, respectively, are shown in Fig. 5.15. In this work, the inlet temperature was 20°C and little pressure drop occurred between inlet and outlet.

The electrolyte temperature rise, measured at inlet and outlet points, shows good agreement with the increase calculated from ohmic resistance heating theory [Equation (5.48)]. However, the temperatures measured within the electrode gap, just before outlet (6 mm), are seen to be greater than those at outlet. As the flow-rate is reduced, the temperature within the gap increases to a maximum value which corresponds closely to the boiling state of the electrolyte. Fluctuations in the machining current and voltage were observed for this condition, and the more the flow-rate was reduced, the further upstream occurred the point of onset of the boiling condition.

The higher temperatures within the gap are consistent with measurements carried out in a region of increased resistivity. Such local conditions would be provided by a hydrogen bubble–electrolyte mixture adjacent to the cathode in which the temperature of the electrolyte around the bubbles would be higher than that in the bulk of the solution.

On the assumption that this boiling condition in the gap is largely induced by the presence of hydrogen gas, the use of an electrolyte like sodium nitrate should provide some striking differences in behaviour

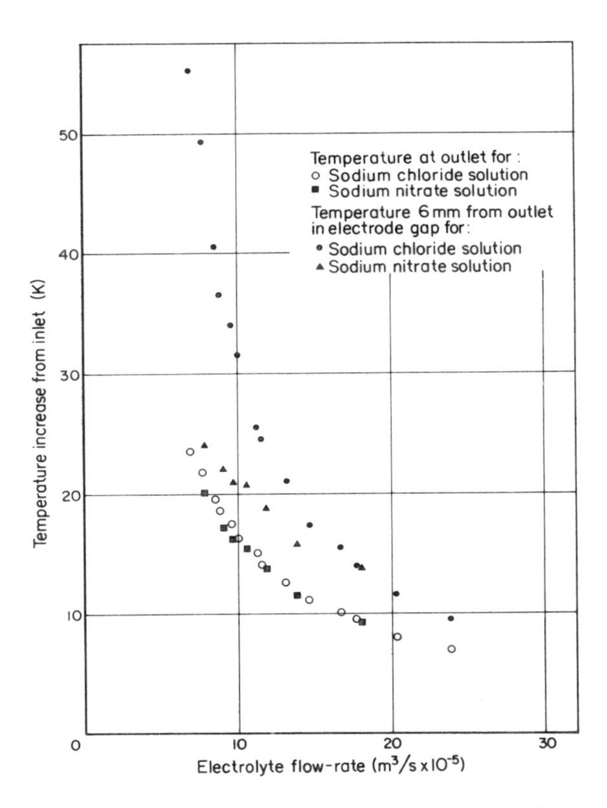

Fig. 5.16 Temperature increase for sodium chloride and sodium nitrate electrolytes (after Clark and McGeough [19])

from those of NaCl solution. It will be recalled from Chapter 4 that mild steel is machined with substantially less hydrogen gas evolution in nitrate electrolytes. Indeed, tests carried out with a 17% (w/w) $NaNO_3$ solution under the same operating conditions did yield quite contrasting results. [A 17% (w/w) $NaNO_3$ solution has the same conductivity as a 10% (w/w) NaCl solution.] In Fig. 5.16, temperatures measured at outlet are comparable with those obtained for the tests with the chloride solution. At high flow-rates, temperatures within the gap are also similar to those in the NaCl test series. At lower flow-rates, however, these temperatures within the gap are lower than those for the equivalent chloride tests. With this reduction in flow-rate there was an associated decrease in current efficiency, from almost 70% to small values (about 2 to 5%), corresponding to conditions of virtually complete passivation, at the highest and lowest flow-rates respectively. (Current efficiencies were based on divalent metal removal.) At virtual passivation conditions, the temperature within the gap is only slightly greater than that at outlet. This observation may be related to the work of Mao who found that no hydrogen was evolved at the cathode under conditions of low current efficiency when machining mild steel with nitrate solutions (see Chapter 4). Three distinct trends in variation of gap width were also determined from these experiments. For current efficiencies close to 70%, the electrode gap increased slightly along the electrode length. As the current efficiency decreased with flow-rate, the gap became gradually convergent; whilst for conditions of virtually complete passivation the gap remained constant along the electrode length.

These experiments also bring to light two conditions which can limit the rate of machining: boiling of the electrolyte and passivation. These conditions, and others which impose limitations on the rate of ECM, are discussed in the next section.

5.9 Limitations on the rate of electrochemical machining

5.9.1 Electrolyte boiling

In Chapter 1, a simple calculation was carried out to estimate the electrolyte flow-rate required to prevent the electrolyte boiling at the gap exit. The study of boiling phenomena in the section above demonstrated that boiling must be prevented throughout the length of the gap, as well as at outlet. A further restraint, accordingly, has to be imposed on any analysis which defines the conditions for the prevention of boiling. In terms of the theory of temperature distribution in the previous section, even a simple condition for no boiling would now become

$$(T_b - T_0) > \left(\frac{V}{e_a c_e}\right)\left(\frac{x}{h_0}\right)\left(\frac{\rho_a}{\rho_0}\right)\left(\frac{f}{U_0}\right)$$

where x is now evaluated at some point along the electrode length at which the electrolyte first boils at temperature T_b. The above condition still indicates, though, that if the velocity, U_0, is increased, boiling will occur further downstream, and that the boiling state will be advanced if the applied voltage and feed-rate are increased.

From Section 5.8, it is again noted that, with the onset of boiling, violent fluctuations in the cell current and voltage are likely to arise.

5.9.2 Sparking

The rate of machining can also be limited by sparking; with its onset, machining is usually terminated. A further effect is often electrode wear. (In practice, possible damage to the cathode can be diminished by the use of materials like stainless steel and tungsten alloys.) Sparking and electrical breakdown in ECM appear to be associated with the formation of a gas blanket over one of the electrodes. Its occurrence is also greatly influenced by the electrolyte velocity; if the flow-rate is sufficiently high, sparking is usually prevented. However, if the flow-rate is reduced, large voltage fluctuations, and eventually sparking, occur. Related oscillations in the cathode potential and cell voltage usually indicate that the phenomena are cathodic in origin [5, 8]. It has also been claimed that their onset coincides with the appearance of a different type of gas bubble, which is large and irregular in shape. This type of formation replaces the more frequent, small individual gas bubbles. The new continuous gas pocket may also cover a large part of the cathode surface, and even if it only instantaneously covers the complete surface, sparking results. The volume fraction at which sparking occurs has been estimated to be about 0·2 at 100 and 150 A/cm^2, the flow-rate being 5 to 10 m/s (equivalent to 20 to 40 x 10^{-6} m^3/s).

If, owing to the presence of non-conducting gas bubbles, the gap develops a sufficiently steep taper, direct physical contact, and sparking, may ensue between anode and cathode. The tapering of the gap was discussed in Section 5.5. It was seen then that this problem can be diminished by use of higher flow-rates.

5.9.3 Cavitation

The presence of cavitation bubbles within the machining gap may also cause machining to be terminated, since these bubbles form non-conducting regions. Moreover, cavitation has been a suggested cause of some very rough, striated finishes in ECM. Clearly, surface roughness of sufficient magnitude can also lead to direct contact between anode and cathode causing an electrical short-circuit.

The simplest means of overcoming cavitation is by the application of an outlet pressure. This procedure has been discussed in Section 2.10.3.

Limitations on the electrode feed-rate due to cavitation have been the subject of a detailed theoretical study of electrolyte flow between shallow, axially symmetric cavities [11]. For a flat-bottomed cathode and for flow in the outward direction, an analysis of the transport equation for momentum, with use of typical numerical values has yielded the set of curves given in Fig. 5.17. Each curve was computed for an absolute pressure at inlet to the machining gap, in a range from about 0·06 times to twice atmospheric pressure. The absolute pressure at outlet to the gap was also about twice atmospheric, and the inlet temperature was 38°C. Since the saturation pressure corresponding to this temperature is approximately 0·06 times atmospheric value, curve (a) indicates a cavitation limit. That is, cavitation would occur at the gap inlet for feed-rate/flow-rate com-

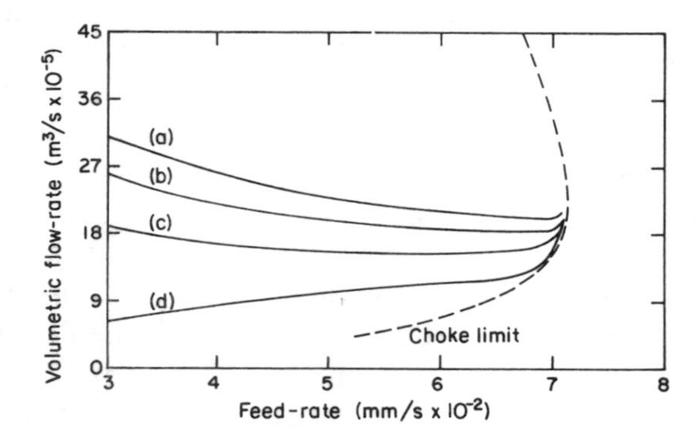

Fig. 5.17 Cavitation limitations on feed-rate. Absolute pressure (atm.) at gap inlet: (a) 0·068, (b) 0·68, (c) 1·36, (d) 2·04 (after Thorpe and Zerkle [11])

binations which lie above the curve (a) limit. ECM is also impossible beyond the broken line on the right of Fig. 5.17. This limitation is imposed by choking. It is discussed more fully below, in relation to flow down a rectangular channel.

5.9.4 Choking

In Chapter 2, the basic principles of choking were outlined. This phenomenon is often encountered in compressible fluid flow or in two-phase fluid flow without ECM. The presence of compressible hydrogen gas in ECM flow means that choking is another possible limitation on the rate of machining. Although no reports are yet available of choking observed in ECM, a theoretical analysis provides some insight into the limiting physical conditions which are imposed by the onset of choking [10].

Because this analysis is rather lengthy, only its main points are summarised here. Essentially, the analysis consists of an examination of the transport equation of motion for flow along a rectilinear gap in ECM. By use of the assumed equation of state, (5.28), a differential equation is obtained for the variation along the electrode length of the hydrogen gas density:

$$\frac{dA}{dx^*} = \frac{\dfrac{CT_r^0 Z^3}{2A} - BY^{2n+1} + A(n+1)Y^{2n}Z + \left(\dfrac{SMK_0}{4}\right)Y^{3n+2}}{\dfrac{C(1+T_r^0 x^*)Z^3}{2A^2} - (n+1)Y^{2n}Zx^*} \qquad (5.58)$$

Here the gas density is expressed in terms of the function,

$$A = \frac{\rho_0 e_g}{\sigma \rho_g e_a} \qquad (5.40)$$

In addition,

$$x^* = \left(\frac{x}{h_0}\right)\left(\frac{\rho_a}{\rho_0}\right)\left(\frac{f}{U_0}\right) \qquad (5.42)$$

$$C = \frac{2}{\sigma}\left(\frac{e_g}{e_a}\right)\left(\frac{RT_0}{U_0^2}\right)$$

$$T_r = \frac{V}{e_a c_e}$$

$$Z = (1 + Bx^*)$$

$$Y = (1 + Ax^*)$$

where $B = \zeta T_r$.

Here M is the frictional pressure drop multiplier for two-phase flow. K_0 is the pressure loss coefficient for a rectilinear gap without ECM.

We recollect that the superscript 0 denotes a dimensionless quantity, and that

$$S = \frac{\rho_0 h_0 U_0}{\rho_a L f}$$

The important point about Equation (5.58) is the possibility of a singularity. That is, as x^* increases from zero value at the gap inlet, the denominator will eventually become zero. This condition corresponds to 'choking'. It will become clear below that the choking condition at the gap exit corresponds to a limiting value of the cathode feed-rate. If choking is to be avoided, the denominator must not become zero for any position along the gap. In particular, at the gap exit, where $x^* = 1/S$, we must have

$$\frac{C\left(1 + T_r^0 \frac{1}{S}\right)\left(1 + B\frac{1}{S}\right)^2}{2A_e^2} > \frac{(n+1)\left(1 + A_e \frac{1}{S}\right)^{2n}}{S} \tag{5.59}$$

It can also be shown that, at the gap exit,

$$A_e = \frac{C}{p_e^0}\left(1 + T_r^0 \frac{1}{S}\right) \tag{5.60}$$

Here the subscript e denotes the exit condition. When Equation (5.60) is substituted into Equation (5.59), the following expression is obtained

$$\frac{2(n+1)C(S + T_r^0)[S^2 p_e^0 + C(S + T_r^0)]^{2n}}{(S+B)^2 p_e^{0\,2(n+1)} S^{4n}} < 1 \tag{5.61}$$

If all process variables, except gap width h_0, electrolyte velocity at inlet U_0, electrolyte pressure p_e, and index n are known, then the constants C, T_r^0, and B can be computed. Since S is proportional to $U_0 h_0^2$, and p_e^0 to p_e/U_0^2, Equation (5.61) can be evaluated for different combinations of the quantities h_0, U_0, p_e, and n. Thorpe and Zerkle show that, if choking is to be avoided, only a limited range of inlet velocities U_0 is available for each h_0, p_e, and n. Some of their results are shown in Fig. 5.18. For outlet pressures of 101 and 102 kN/m^2 and n values of 1, 1.5, and 2, acceptable values of the gap width at inlet, h_0, are to the right of each curve shown. Since the feed-rate f is inversely proportional to h_0, the exit pressure and n have a significant influence on the maximum feed-rate.

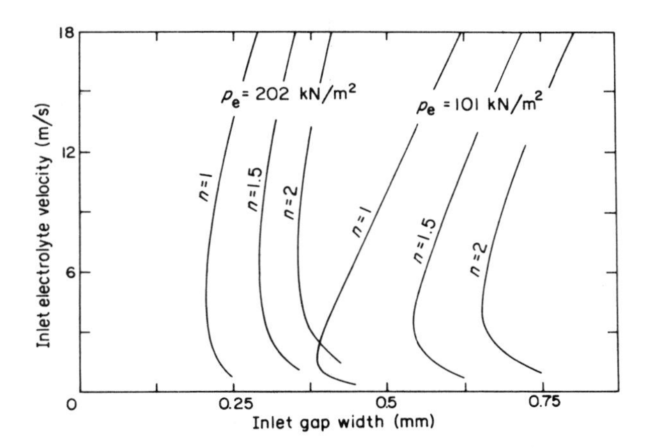

Fig. 5.18 Limitations imposed by choking on inlet flow velocity, U_0, and inlet gap width, h_0; $\kappa_0 = 0{\cdot}2\ \text{ohm}^{-1}\ \text{cm}^{-1}$; $L = 25{\cdot}4$ mm; $T_0 = 20°\text{C}$, $e_a = 29 \times 10^{-5}$ g/C; $e_g = 10^{-5}$ g/C; $\nu = 1\ \text{mm}^2/\text{s}$; $\sigma = 1$; $\rho_a = 7{\cdot}8\ \text{g/cm}^3$; $\rho_0 = 1\ \text{g/cm}^3$ (after Thorpe and Zerkle [10])

5.9.5 *Passivation*

In the course of Chapter 4 many examples were given of the limitations on the rate of ECM arising from a passivating film on the anode surface. Even in Section 5.8, passivation was shown to impose considerable restrictions on the machining of mild steel in sodium nitrate solution. From that study, one other result is of interest: a computation of the Nusselt number for the passivation conditions [17]. (The basis of this calculation is also outlined in Chapter 4.)

The Nusselt number is

$$\text{Nu} = \frac{J_l d_h}{zFC_b D}$$

where J_l is the limiting current density, d_h the hydraulic mean diameter, z the valency of the dissolving metal ions, F the Faraday, C_b the bulk concentration of the reacting species, and D the diffusion coefficient.

Suppose that J_l is determined from the experimentally observed current density at passivation (16 A/cm^2), and C_b is replaced by a possible saturation concentration value (say, 5 mole/litre). On substitution of these values together with d_h = 4 mm, z = 2, F = 96 500 C $D = 1{\cdot}5 \times 10^{-5}$ cm^2/s, the Nusselt number is calculated to be about 440.

Although this estimate indicates the order of magnitude of the Nusselt number, the value of the number would clearly be changed on the substitution of some other possible saturation concentration.

5.9.6 Limiting current densities for ionic mass transport

The maintenance of the electrode reactions in ECM requires that the anodic and cathodic products be removed from the neighbourhood of the electrodes and that reactants be introduced. The movement of the dissolved particles has been shown in Chapter 3 to be describable in terms of the Nusselt number, with which are usually associated the Reynolds and Schmidt numbers (Equations (3.32) to (3.38)).

Landolt et al. [20] suggest that these relationships can be used to predict limiting current densities in ECM. They point out that, since the anodic dissolution products have to be removed from the anode region, the dissolution rate may be checked by the limit of solubility of the reaction products. They advocate, therefore, that the C_b term in these expressions be replaced by a term, ΔC, for the concentration difference between the interface and the bulk solution which is a measure of the solubility. For example, $\Delta C = 0{\cdot}1$ to 10 mole/litre could represent concentration differences ranging from weakly to well soluble conditions. Clearly, the limiting current densities in conditions for which ΔC = 10 mole/litre will be higher than those for which $\Delta C = 0{\cdot}1$ mole/litre.

References

1. McGeough, J. A. and Rasmussen, H., *Trans. A.S.M.E. Series B, J. Eng. Ind.* (1970) No. 2, 400.
2. Johnson, M. P. and Brown, D. J., Dept. Trade & Ind. Report, Contract No. KJ/4M/117/CB78A (1970) July.
3. Ito, S., Chikamori, K. and Hayashi, T., *J. Mech. Lab. Japan* (1964) **10**, No. 2, 33.
4. Noble, C. F. and Shine, S. J., Paper presented at First Int. Conf. on ECM, Leicester University, Mar. 1973.

5. Landolt, D., Acosta, R., Muller, R. H. and Tobias, C. W., *J. Electrochem. Soc.* (1970) **117**, No. 6, 839.
6. Hopenfeld, J. and Cole, R. R., *Trans. A.S.M.E., J. Eng. Ind.* (1966) **88**, No. 4, 455.
7. Hopenfeld, J. and Cole, R. R., *Trans. A.S.M.E., J. Eng. Ind.* (1969) Aug., No. 8, 755.
8. Baxter, A. C., Ph.D. Thesis, University of Warwick (1967).
9. Baxter, A. C., Freer, H. E. and Willenbruch, D. A., Paper presented at First Int. Conf. on ECM, Leicester Univ., Mar. 1973.
10. Thorpe, J. F. and Zerkle, R. D., *Int. J. Mach. Tool Des. Res.* (1969) **9**, 131.
11. Thorpe, J. F. and Zerkle, R. D., in *Fundamentals of ECM* (ed. C. L. Faust), see Ref. 8, Ch. 4, 1.
12. Thorpe, J. F., Paper presented at Soc. Manuf. Eng. Conf., Detroit, 1970, Paper No. MR-70-513.
13. Bass, L., *Trans. Faraday Soc.* (1964) **60**, 1646.
14. De la Rue, R. E. and Tobias, C. W., *J. Electrochem. Soc.* (1959) Sept, 827.
15. Tipton, H., Proc. Fifth Int. Conf. Mach. Tool Res., Birmingham (1964) Sept., p. 509.
16. Clark, W. G., McGeough, J. A. and Yeo, J. T., unpublished work.
17. Clark, W. G., Ph.D. Thesis, Strathclyde University, Glasgow (to be submitted).
18. Tobias, C. W., Paper presented at First Int. Conf. on ECM, Leicester University, Mar. 1973.
19. Clark, W. G. and McGeough, J. A., Paper presented at First Int. Conf. on ECM, Leicester University, Mar. 1973.
20. Landolt, D., Muller, R. H. and Tobias, C. W., in *Fundamentals of ECM* (ed. C. L. Faust), see Ref. 8, Ch. 4, p. 200.

CHAPTER SIX

Smoothing of an Irregular Anode Surface

If a particular cathode shape is chosen for an ECM operation, the anode form should eventually become approximately complementary to it. Studies of the shaping process of ECM are often concerned with the variation in the shape of the anode as machining proceeds, and with the machining time required to achieve the steady-state shape on the anode. They should also deal with the influence of overpotentials, and with the effects of the electrolyte flow. (The question of flow has yet to be tackled in any definitive fashion, and is not considered in this chapter.)

The simplest case of shaping which allows these problems to be studied is anodic smoothing. Here the cathode has a flat plane face, and the anode surface initially carries surface irregularities. These are gradually removed by ECM, so that eventually the anode surface also becomes flat, and then, of course, resembles the cathode shape. The practical analogy of this problem is electrochemical deburring, in which surface irregularities are removed electrochemically from an anode workpiece.

6.1 Basic equations

Three basic equations are used:

(i) Laplace's equation

$$\nabla^2 \phi = 0 \tag{6.1}$$

the solution of which will give the potential ϕ at any point in the electrolyte, particularly at the electrode surfaces.

(ii) Ohm's law

$$J = -\kappa_e \nabla \phi \tag{6.2}$$

where J, the current density, will be found from the potential known from (6.1), κ_e being the electrolyte conductivity.

(iii) Faraday's law

$$\dot{r}_a = \frac{AJ}{zF\rho_a} \tag{6.3}$$

which will be used to give the anode recession rate $\dot{r}$, A being the atomic weight, z the valency, ρ_a the density of the anode metal, and F is Faraday's constant.

The conditions under which these equations can be used – the 'quasi-steady model' of the ECM process – have been investigated in some detail [1, 2]. The boundary conditions for the potential at the electrodes, and the conditions under which Ohm's law and Faraday's law are applicable, are now discussed.

6.2 Potential boundary conditions

If the electrode surfaces are equipotentials, then the boundary conditions are

$$\phi = 0 \text{ at the cathode} \tag{6.4}$$

$$\phi = V \text{ at the anode} \tag{6.5}$$

where V is the applied potential difference. However, we have seen in Chapter 3 that the reactions at both electrodes cause current density-dependent overpotentials. Their presence at the electrodes alters the boundary conditions to

$$\phi = f(J) \text{ at the cathode}$$

$$\phi = V - g(J) \text{ at the anode}$$

where $f(J)$ and $g(J)$ are arbitrary functions for the cathodic and anodic overpotentials respectively.

6.3 Applicability of Ohm's law and Faraday's law

In Chapter 3, we saw that the potential drop across the diffusion layer, associated with the concentration overpotential, is ohmic. In the bulk electrolyte, outside the diffusion layer, all concentration gradients can be assumed to be destroyed by the agitation of the electrolyte. Accordingly, Ohm's law can be applied in the form $\boldsymbol{J} = \kappa_e \boldsymbol{E}$ where κ_e is the bulk conductivity and $\boldsymbol{E}$ is the electric field. If the diffusion layer thickness and potential drop across it are assumed to be sufficiently small, the conductivity can be assumed to have its bulk value everywhere. The conductivity is also assumed to remain constant everywhere. Joule heating and the formation of hydrogen gas bubbles which, as we have seen, respectively increase and decrease the effective conductivity are assumed to be suppressed by sufficient agitation of the electrolyte.

The use of Faraday's law implies that all the current at the anode is used to dissolve the metal; that is, no other reaction (e.g. oxygen evolution) occurs there.

With these conditions and assumptions, certain solutions of the electrochemical shaping problem are possible. These involve the solution of an elliptic differential equation with moving boundaries. In general, this is difficult; however, if the boundaries are assumed to change so slowly that their motion may be ignored at any given time, then the equation may be solved for any instant of time. From this solution, further equations can be derived which describe the behaviour of the boundary over the next small period of time, when Laplace's equation may again be solved, and so on.

The most tractable cases give rise to boundary-describing equations which are 'self-similar' [3]. These do not involve time explicitly, and may be transformed into equations which do not depend explicitly on the shape of the anode surface. Such cases are considered for

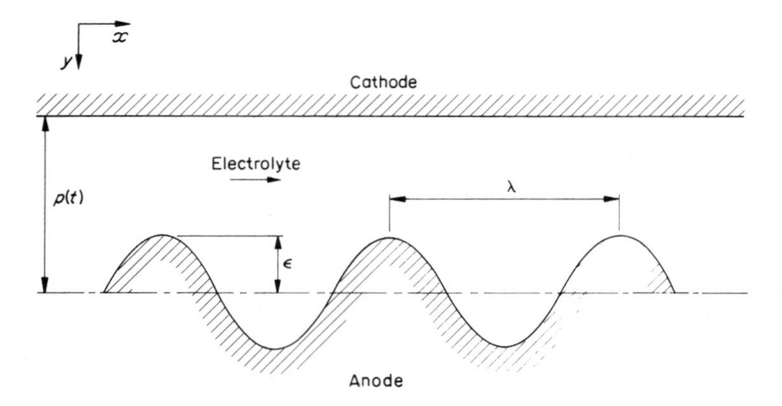

Fig. 6.1 Configuration of electrodes

two-dimensional forms, although the extension to three dimensions will also be outlined. The work will be developed initially without the complication of overpotential effects, i.e. at the cathode a boundary condition $\phi = 0$, and at the anode $\phi = V$, will be assumed. After the physical and mathematical principles have been fully considered, the effect of overpotentials will be included.

6.4 The basic problem

We investigate the smoothing of small irregularities on the anode surface, the configuration of the electrodes being shown in Fig. 6.1. Sinusoidal irregularities of wavelength $\lambda(= 1/k$, k being the wave number) are considered.

Suppose the plane cathode is given by

$$y = 0 \tag{6.6}$$

and the anode by

$$y = p + \epsilon \sin kx \tag{6.7}$$

where the quantity p is the 'average gap' between the electrodes. The gross behaviour of the anode surface will later be determined by this parameter. The maximum amplitude of the irregularities is given by ϵ, the initial value being $\epsilon(0)$. Initial values of p and ϵ are assumed known; we investigate their subsequent behaviour. It is assumed that

$$\epsilon \ll p \tag{6.8}$$

Initially a solution is sought for Laplace's equation in two dimensions:

$$\frac{\partial^2 \phi}{\partial x^2} + \frac{\partial^2 \phi}{\partial y^2} = 0 \tag{6.9}$$

with the boundary conditions

$$\phi = 0 \quad \text{on } y = 0 \tag{6.10}$$

$$\phi = V \quad \text{on } y = p + \epsilon \sin kx \tag{6.11}$$

Expansion of the solution in terms of the small parameter $\epsilon(0)/p$ gives (to first order in $\epsilon(0)/p$):

$$\phi = \phi_0 + \frac{\epsilon(0)}{p}\phi_1 \tag{6.12}$$

where ϕ_0 is the potential between two plane electrodes at a distance p apart, and ϕ_1 is a non-singular perturbation. The potential ϕ_0 satisfies Laplace's equation, with the boundary conditions

$$\phi_0 = 0 \quad \text{on } y = 0 \tag{6.13}$$

$$\phi_0 = V \quad \text{on } y = p \tag{6.14}$$

The solution is

$$\phi_0 = \frac{Vy}{p} \tag{6.15}$$

ϕ_1 is now chosen to satisfy Laplace's equation, with the boundary conditions

$$\phi_1 = 0 \quad \text{on } y = 0 \tag{6.16}$$

$$\phi_0 + \frac{\epsilon(0)}{p}\phi_1 = V \quad \text{on } y = p + \epsilon \sin kx \tag{6.17}$$

Now, on $y = p + \epsilon \sin kx$,

$$\phi_0 = V + \frac{V\epsilon}{p} \sin kx \tag{6.18}$$

the second boundary condition becomes

$$\phi_1 = -\frac{V\epsilon}{\epsilon(0)} \sin kx \tag{6.19}$$

Separation of variables and application of the first boundary condition give

$$\phi_1 = A \sin kx \sinh ky \tag{6.20}$$

where the constant A is to be determined. Applying the second boundary condition (6.19), we obtain, to first-order approximation in $\epsilon(0)/p$:

$$A = -\frac{V}{\sinh kp}\left(\frac{\epsilon}{\epsilon(0)}\right) \tag{6.21}$$

From Equations (6.12), (6.15), and (6.21), the expression for the potential is obtained:

$$\phi = \frac{Vy}{p} - \frac{V\epsilon}{p}\frac{\sin kx \sinh ky}{\sinh kp} \tag{6.22}$$

6.5 Behaviour of anode surface and of surface irregularities

By the use of this expression for ϕ, a set of equations can be derived which describes separately the behaviour of the anode surface and the behaviour of the irregularities. On the anode surface, consider $\mathrm{d}y_a/\mathrm{d}t$, the reduction rate in the vertical direction. (Here, and subsequently, the subscript denotes evaluation at the anode surface.) The field in this direction is $-\partial\phi/\mathrm{d}y$. If the anode surface is described by a vector function $\boldsymbol{r}_a(x,y)$, then, from Ohm's law and Faraday's law, we obtain

$$\frac{\mathrm{d}\boldsymbol{r}_a}{\mathrm{d}t} = M\nabla\phi_a \tag{6.23}$$

where M ($= A\kappa_e/Fz\rho_a$, ρ_a being the anode metal density) is a constant. The vertical component of Equation (6.23) is

$$\frac{\mathrm{d}y_a}{\mathrm{d}t} = M\frac{\partial\phi_a}{\partial y} \tag{6.24}$$

The field at the top and foot of the irregularities is now examined. At the top, $x = (\pi/k)(2n + 3/2)$ with $n = 0, 1, 2, \ldots; y = p - \epsilon$. Here, the field has no x component, and is given by

$$\begin{aligned}
\frac{\partial\phi}{\partial y} &= \frac{V}{p}\,\frac{V\epsilon k}{p}\,\frac{\sin(3\pi/2)\cosh k(p-\epsilon)}{\sinh kp} \\
&\simeq \frac{V}{p}\left[1 + \epsilon k\left(\frac{\cosh kp\cosh k\epsilon}{\sinh kp} + \frac{\sinh kp\sinh k\epsilon}{\sinh kp}\right)\right] \\
&\simeq \frac{V}{p}(1 + \epsilon k\coth kp)
\end{aligned}$$

Neglected, as before, are terms of order higher than the first in ϵ/p.

Next, Equation (6.24) is used to calculate the amount of metal removed from the top and foot of the irregularity. In time $\mathrm{d}t$ the top of the irregularity will have been lowered by an amount $\mathrm{d}(p - \epsilon)$, where

$$\mathrm{d}(p-\epsilon) = \frac{MV}{p}(1 + \epsilon k\coth kp)\mathrm{d}t \tag{6.25}$$

Similarly, the reduction in height of the foot of the irregularity in time dt is given by

$$\mathrm{d}(p + \epsilon) = \frac{MV}{p}(1 - \epsilon k \coth kp)\mathrm{d}t \tag{6.26}$$

The average of these two quantities is the reduction in height of the average anode surface, causing an increase in the average gap due to ECM. This reduction is offset by a decrease in the average gap due to the forward movement of the cathode, assumed constant at rate f. Using Equations (6.25) and (6.26), we obtain

$$\frac{\mathrm{d}p}{\mathrm{d}t} = \frac{MV}{p} - f \tag{6.27}$$

in agreement with elementary ohmic theory for two plane electrodes at the average gap [see Equation (5.3)].

The difference between the quantities obtained in Equations (6.25) and (6.26) divided by 2, is the decrease in height of the irregularity in time dt; we obtain

$$\frac{\mathrm{d}\epsilon}{\mathrm{d}t} = -\frac{MVk}{p} \coth kp\ \epsilon \tag{6.28}$$

as the differential equation describing the time behaviour of the height of the irregularities.

We have thus derived, from our solution of Laplace's equation, a set of equations describing separately the gross behaviour of the anode surface and the behaviour of the irregularities. Solving Equation (6.27), we obtain, as usual;

$$t = \frac{1}{f}\left[p(0) - p + \frac{MV}{f} \ln \frac{MV - fp(0)}{MV - fp}\right] \tag{6.29}$$

which may be expressed as

$$t = \frac{1}{f}\left[p(0) - p + p_{\mathrm{e}} \ln \frac{p_{\mathrm{e}} - p(0)}{p_{\mathrm{e}} - p}\right] \tag{6.29a}$$

where $p(0)$ is the initial gap, and $p_{\mathrm{e}} = MV/f$ is the equilibrium gap for the condition specified. Equation (6.29) gives p as an implicit function of t.

Equation (6.27) is quite basic, and has been derived in Chapter 5. Briefly, we recollect that, if $p(0)$ is greater than p_{e}, then p is always greater than p_e, and tends to it asymptotically, and vice versa if $p(0)$ is less than p_e.

From Equation (6.28), on transformation:

$$\frac{d\epsilon}{dp} = \frac{d\epsilon}{dt}\frac{dt}{dp} = -\frac{MVk}{MV - fp} \coth kp\, \epsilon$$

which yields

$$\epsilon = \epsilon(0) \exp\left[-MVk \int_{p(0)}^{p} \frac{\coth ks}{MV - fs} ds\right] \qquad (6.30)$$

provided that dp/dt is not zero, i.e. the process is not being carried out at the equilibrium gap. The formula (6.30) gives ϵ as a function of p, and hence implicitly as a function of time. Note that, as expected ϵ decreases, whether p is increasing or decreasing to its equilibrium value.

Two cases of interest are tractable, depending on the value of the parameter kp.

6.5.1 *Short-wavelength irregularities*

For the short-wavelength case ($kp \gg 1$), the wavelength is small compared with the gap width. Since $\coth kp$ can be taken to be unity, Equation (6.30) becomes

$$\epsilon(p) = \epsilon(0)\left(\frac{MV - fp(0)}{MV - fp}\right)^{-MVk/f} \qquad (6.31)$$

provided that p does not assume its equilibrium value MV/f. Here the dependence of ϵ on p is a power of the quantity $[MV - fp(0)]/(MV - fp)$, and the index depends linearly on the wave number, i.e. the shorter the wavelength, the more quickly are the irregularities dissolved. The significance is discussed more fully in the next section.

6.5.2 *Long-wavelength irregularities*

In the long-wavelength case ($kp \ll 1$), Equation (6.30) gives

$$\epsilon = \epsilon(0)\frac{p(0)(MV - fp)}{p[MV - fp(0)]} \qquad (6.32)$$

(again with the condition $p \neq MV/f$). This is identical with the simple formula for the reduction of a single step on the anode (i.e. assuming that the current travels only normally between the electrodes). The result is expected, since, with long-wavelength irregularities, the crests and troughs may be considered to be sections of plane electrodes at different distances from the cathode (Fig. 6.2). The similarity is also discussed more fully below.

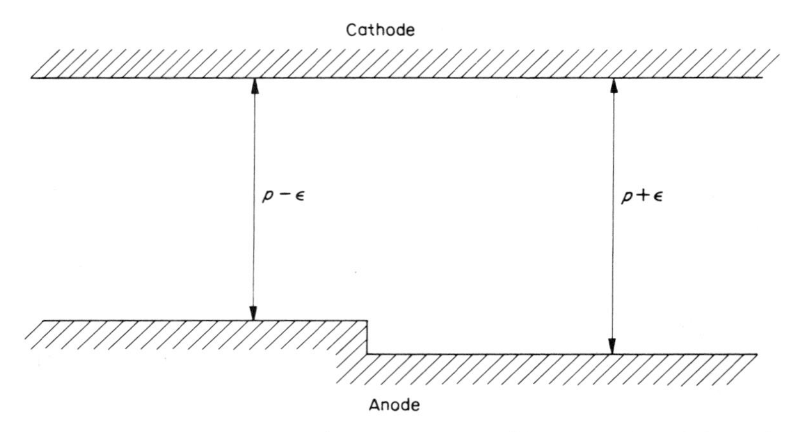

Fig. 6.2 Simple stepped anode (approximation of long-wavelength irregularities; Section 6.5.2)

For equilibrium gap conditions, the solutions of Equations (6.27) and (6.28) are

$$p = \frac{MV}{f} \tag{6.33}$$

and

$$\epsilon = \epsilon(0) \exp\left[-\left(\frac{MVk}{p} \coth kp\right)t\right] \tag{6.34}$$

In later examples, such gap conditions are used.

6.6 Field concentration effects

In a simple theory the field lines across the gap may be considered to be parallel, and the rate of reduction of the irregularities may be obtained by an ohmic theory based on the differences in gap size between corresponding points on the electrodes. The present theory accounts for the field concentration effects: the field at a point where the curvature of the electrode surface is large and convex is much greater than the field at a plane surface. Since short-wavelength irregularities have a much greater concentration of field at their crests than longer wavelengths, and thus have a greater difference in field between the peak and foot of a wave, more rapid smoothing is to be expected. This has been shown above.

For long-wavelength irregularities, $d\epsilon/dt$ may be found as a first approximation from Ohm's law, assuming a parallel field distribution. At the top of the wave, the field is $V/(p-\epsilon)$, and at the foot it is $V/(p+\epsilon)$. The rate of reduction of the furrow size is

$$\frac{d(2\epsilon)}{dt} = -M\left(\frac{V}{p-\epsilon} - \frac{V}{p+\epsilon}\right)$$

$$\frac{d\epsilon}{dt} = -\frac{MV\epsilon}{p^2} \tag{6.35}$$

Comparing Equation (6.35) with (6.28) for the sinusoidal surface, we see that, when the wavelength is large compared to the gap ($kp \ll 1$, and $\coth kp \simeq 1/kp$), Equation (6.28) reduces to (6.35), i.e. the differential equations derived in the present theory reduce to the simple ohmic equations when the conditions justify the use of the approximations of parallel field theory.

Consider Equation (6.28) when the wavelength is moderately large compared with p. Then

$$\coth kp \simeq \frac{1}{kp} + \frac{kp}{3}$$

Equation (6.28) becomes

$$\frac{d\epsilon}{dt} = -MV\left(\frac{1}{p^2} + \frac{k^2}{3}\right)\epsilon$$

When the gap remains constant, integration gives

$$\epsilon = \epsilon(0)\exp\left(-\frac{MV}{p^2} - \frac{MVk^2}{3}\right)t$$

The term $\exp[-(MVk^2/3)t]$ may be regarded as a first field-concentration correction. In this case the wavelength, while still large, is sufficiently small to cause an appreciable field-concentration effect. Note that this term is independent of p, i.e. the increase in the smoothing rate, due to field concentration, surprisingly is independent of the gap width.

6.7 Surface smoothing to a required tolerance

This work has provided further information about a feature of electrochemical deburring [4]. With this process, it is useful to know the overall depth of metal which has to be machined to permit the removal of a surface irregularity, the so-called 'memory

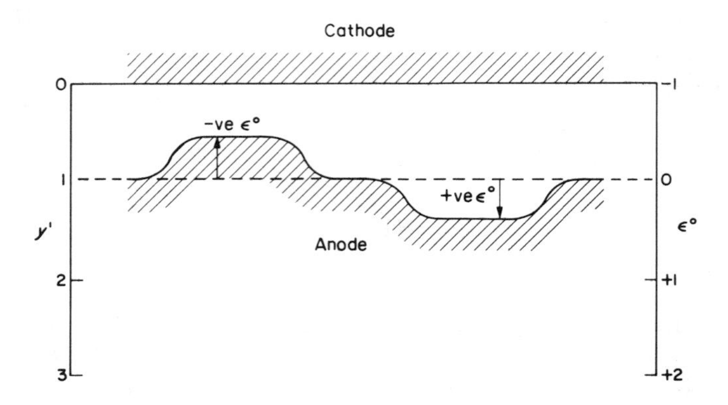

Fig. 6.3 Anode irregularities (after Tipton [5])

effect'. One of the first reports of studies on this topic is that by Tipton [5].

In this investigation, a plane cathode is again used to produce a plane surface on the anode. The latter electrode is assumed to carry initially irregularities whose surfaces are plane-faced (Fig. 6.3). That figure shows two such anodic irregularities, which are at different distances from the cathode, and assumed to be sufficiently distant from each other. Then, simple ohmic theory, incorporating parallel field lines across the gap, can be applied separately to each irregularity. As ECM proceeds, the distance from the cathode of each region of irregularity will tend to the equilibrium value. This equilibrium distance is, of course, $MV/f = h_e$. If we work in terms of dimensionless quantities, h^0 and t^0, where

$$h^0 = \frac{fh}{MV} = \frac{h}{h_e} \tag{6.36}$$

and t^0 is the time taken to move one equilibrium gap distance:

$$t^0 = \frac{f^2 t}{MV} \tag{6.37}$$

The usual equation for the gap now becomes

$$t^0 = h^0(0) - h^0 + \ln \frac{h^0(0) - 1}{h^0 - 1} \tag{6.38}$$

The steady-state position, $h^0 = 1$, can now be considered to be the final anode surface determined by the cathode surface. The deviations from this ideal surface by the irregularities can now be expressed non-dimensionally

$$\epsilon^0 = h^0 - 1$$

Clearly, the above equation (6.38) for h^0 can be expressed in terms of ϵ^0:

$$t^0 = \epsilon^0(0) - \epsilon^0 + \ln \frac{\epsilon^0(0)}{\epsilon^0} \tag{6.39}$$

Next, the time required to machine the anode to a required accuracy, or tolerance, can be found. For instance, suppose that the equilibrium gap is 0·5 mm and that the required tolerance is 0·05 mm. In units of the equilibrium gap, this tolerance is 0·1. The time needed to machine sufficient depth of metal to reduce an irregularity of initial size $\epsilon^0(0)$ to a required size ϵ^0 can be calculated directly from Equation (6.39). Figure 6.4 shows results for small initial irregularities (in terms of equilibrium gaps). The required depth of machining is seen to be very dependent on the required tolerance. For large initial irregularities, the height of the irregularity must be machined; in addition, a further small depth of metal, equal to $\ln[\epsilon^0(0)/\epsilon^0] - \epsilon^0$,

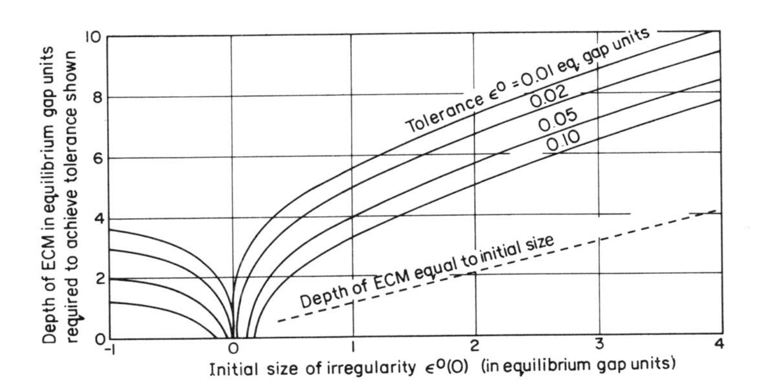

Fig. 6.4 Depth of ECM necessary to reduce small initial defects of size $\epsilon^0(0)$ to required tolerance (after Tipton [5])

must be machined. In Fig. 6.5 this additional depth of metal which must be machined for large heights of irregularities is also presented.

The difficulty of the anodic smoothing of small irregularities to a required tolerance becomes apparent from this analysis. But the

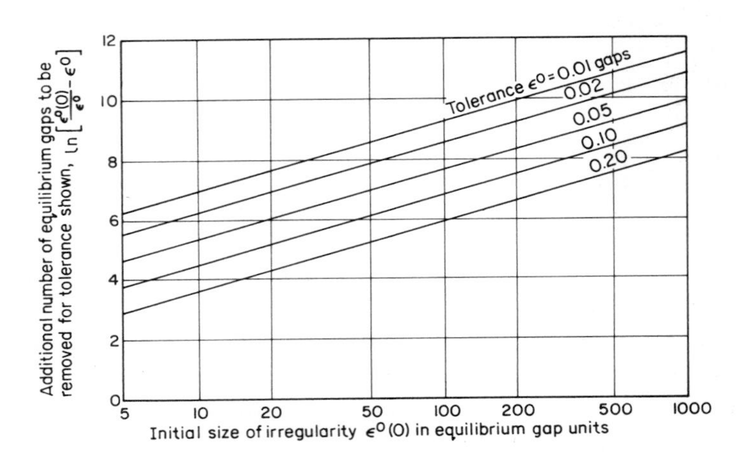

Fig. 6.5 Additional depth of ECM, ln $[\epsilon^0(0)/\epsilon^0 - \epsilon^0]$, in excess of the original defect size necessary to reduce a large initial defect $\epsilon^0(0)$ to a specified tolerance of ϵ^0 (after Tipton [5])

preceding studies indicate a possible means of smoothing to a required tolerance under non-equilibrium ECM conditions so that a minimum amount of metal is lost.

Consider, first, short-wavelength irregularities: $kp \gg 1$ and coth $kp \simeq 1$. If the position of the average anode surface is denoted by the coordinate s, relative to a *fixed* origin, then

$$\frac{\mathrm{d}s}{\mathrm{d}t} = \frac{MV}{p} \tag{6.40}$$

But

$$\frac{\mathrm{d}\epsilon}{\mathrm{d}t} = -\frac{MVk}{p}\epsilon$$

so that

$$\frac{\mathrm{d}s}{\mathrm{d}\epsilon} = -\frac{1}{k\epsilon}$$

That is,

$$s = s(0) + \ln\left[\frac{\epsilon(0)}{\epsilon}\right]^{1/k} \tag{6.41}$$

where $s(0)$ is the initial value of s. Thus s has no explicit dependence on the behaviour of p, and these equations hold for both equilibrium and non-equilibrium conditions. In this case, therefore, surface irregularities cannot be reduced at a rate greater than that achieved in equilibrium conditions. That is, there is no possibility of reducing the loss of the bulk anode metal.

Next, consider the long-wavelength case: $kp \ll 1$ and $\coth kp \simeq 1/kp$. Since

$$\frac{d\epsilon}{dt} = -\frac{MV}{p^2}\epsilon$$

But by means of Equation (6.40),

$$\frac{ds}{d\epsilon} = -\frac{p}{\epsilon}$$

That is,

$$|ds| = p\frac{|d\epsilon|}{\epsilon}$$

For a reduction $d\epsilon$ in the size of the surface irregularities, a corresponding lowering ds of the anode surface has to occur. The smaller the value of the mean gap width, p, the smaller will be this amount ds of bulk anode recession.

6.8 Three-dimensional anode irregularities

These can be treated by a method similar to that given above; the anode surface is described by

$$y_a = p + \epsilon \sin kx \sin lz$$

where l is the wave number of the irregularities in the z direction (perpendicular to the x, y plane). The potential is also deduced by the same procedure:

$$\phi = \frac{Vy}{p} - \frac{V\epsilon}{p}\,\frac{\sin kx \sin lz \sinh[\sqrt{(k^2 + l^2)}y]}{\sinh[\sqrt{(k^2 + l^2)}p]} \tag{6.42}$$

as are the equations for p and ϵ:

$$\frac{dp}{dt} = \frac{MV}{p} - f \tag{6.43}$$

$$\frac{d\epsilon}{dt} = -\frac{MV\sqrt{(l^2 + k^2)}}{p} \coth[\sqrt{(l^2 + k^2)}p]\epsilon \tag{6.44}$$

The behaviour of the anode surface is entirely similar to that of the two-dimensional surface described above. Smoothing here will be more rapid; this is to be expected, since the field concentration at the 'hills' will be greater than that around 'troughs'.

6.9 Arbitrarily shaped irregularities

The theory presented above can be extended, by the use of Fourier series, to cover any shape of anode irregularity. Detailed analysis of this treatment is available elsewhere [3]. For brevity here, only the significant results are quoted. Suppose the irregularities are described by Fourier sine series (although they can be described just as easily by cosine series, or by a combination of sine and cosine series). Then the Fourier coefficients can be shown to behave independently, and the electric field can be obtained by linear superposition of components derived from the individual Fourier terms.

Although the plane cathode is still given by $y = 0$, the anode is now described in terms of a Fourier series:

$$y = p + \epsilon(0) \sum_{1}^{\infty} a_n \sin \frac{n\pi x}{\lambda} \tag{6.45}$$

where 2λ is the fundamental wavelength of the irregularities, and the wave number k has been replaced by $n\pi/\lambda$; a_n is the Fourier coefficient. As usual, we assume that

$$\epsilon = \left[\epsilon(0)^2 \sum_{1}^{\infty} |a_n|^2\right]^{1/2} \ll p$$

where ϵ represents the maximum amplitude of the irregularities. Here $\epsilon(0)$ is a scaling factor for the initial anode shape.

The coefficients a_n are considered to be time-dependent, and differential equations are derived for them. These equations enable description of the time-dependence of the overall height of the irregularities and also of the variation of the shape of the anode surface with machining time.

Analysis, similar to that in Section 6.4, yields the usual expression for the rate of change of gap:

$$\frac{dp}{dt} = \frac{MV}{p} - f$$

and a differential equation for the coefficients a_n:

$$\frac{da_n}{dt} = -\frac{MV\pi}{p\lambda} na_n \coth \frac{n\pi p}{\lambda} \qquad (6.46)$$

Again, as in Section 6.5, and p being assumed constant, two limiting cases can be considered.

(i) Wavelengths small compared with the gap, i.e.

$$\frac{n\pi p}{\lambda} \gg 1, \quad \text{so that } \coth \frac{n\pi p}{\lambda} \simeq 1.$$

Integration of the above equation yields

$$a_n(t) = a_n(0) \exp\left(-\frac{MVn\pi}{p\lambda} t\right) \qquad (6.47)$$

From this result, we deduce that the shape of an arbitrary irregularity becomes that of a sinusoidal one with the fundamental wavelength.

(ii) Wavelengths large compared with the gap, i.e.

$$\frac{n\pi p}{\lambda} \ll 1, \quad \text{so that } \coth \frac{n\pi p}{\lambda} \simeq \frac{\lambda}{n\pi p}$$

On integration of Equation (6.46),

$$a_n(t) = a_n(0) \exp\left(-\frac{MV}{p^2} t\right) \qquad (6.48)$$

This result is the simple ohmic one, previously derived.

6.10 Extension of above theory for even and arbitrarily shaped irregularities

Results similar to those of Section 6.5 can be obtained for irregularities represented by a Fourier cosine series (e.g. cusp-shaped irregularities).

Suppose that the anode is represented by

$$y_a = h(t) - a_0\epsilon(0) - \epsilon(0)\sum_1^\infty a_n(t)\cos\frac{n\pi x}{\lambda} \tag{6.49}$$

Clearly, $p(t)$ is now represented by $h(t) - \epsilon(0)a_0(t)$, and the anode surface is not symmetric about the line defined by the average gap p, as was the case previously. However, similar equations for $a_n(t)$ can be deduced, and for the symmetric case, the expression for $\epsilon(t)$ is

$$\epsilon(t) = \epsilon(0)\sum_1^\infty a_{2n-1}(t) \tag{6.50}$$

When the top and foot of the irregularities do not occur at the positions of symmetry, the anode shape may be represented by a general Fourier series:

$$y_a = p(t) + \epsilon(0)\sum_1^\infty a_n \cos\frac{n\pi x}{\lambda} + \epsilon(0)\sum_1^\infty b_n \sin\frac{n\pi x}{\lambda} \tag{6.51}$$

If the top and foot occur at $x = \alpha$ and $x = \beta$ respectively, the height of the irregularities is

$$\epsilon(t) = \epsilon(0)\sum_1^\infty a_n\left(\cos\frac{n\pi\beta}{\lambda} - \cos\frac{n\pi\alpha}{\lambda}\right) + \epsilon(0)\sum_1^\infty b_n\left(\sin\frac{n\pi\beta}{\lambda} - \sin\frac{n\pi\alpha}{\lambda}\right) \tag{6.52}$$

In general, α and β will be functions of time; this is to be expected, since the shape of the irregularity will change to a sinusoidal one, with the basic wavelength.

Example

Suppose the irregularities consist of a series of rectangular steps, $\epsilon(0)$ being the initial step height. Then

$$a_n(0) = \begin{cases} \dfrac{2}{n\pi}, & n \text{ odd} \\ 0, & n \text{ even} \end{cases}$$

If the basic wavelength 2λ is assumed sufficiently large so that the long-wavelength approximation may be used for each component, then

$$a_n(t) = \frac{2}{\pi n}\exp\left(-\frac{MV}{p}t\right)$$

for n odd. We obtain that (for constant p)

$$\epsilon(t) = \frac{4\epsilon(0)}{\pi} \exp\left(-\frac{MV}{p^2}t\right) \sum_1^\infty (-1)^r \frac{1}{2r+1}$$

$$= \epsilon(0) \exp\left(-\frac{MV}{p^2}t\right)$$

since

$$\sum_1^\infty (-1)^r \frac{1}{2r+1} = \frac{\pi}{4}$$

This is the ohmic result and is to be expected, since a long-wavelength step is a case where the basic ohmic theory will be reasonably accurate.

Figure 6.6 shows how the rectangular step changes with time to a sinusoidal irregularity. Notice here that the theory predicts an initial virtual growth of the anode at the base of the step; this is due to a breakdown of the linearization approximation near areas of abrupt change of anode slope. However, it is shown later that such areas are rapidly removed; thus the above theoretical paradox has negligible bearing on the subsequent changes of the anode profile.

Shapes for which the coefficients of the higher harmonic terms are appreciable in comparison with those of the basic terms will be smoothed rapidly, for the higher harmonic (shorter wavelength) terms are reduced most rapidly. If the effect of shape and spacing of irregularities on smoothing time is investigated, sharp profiles

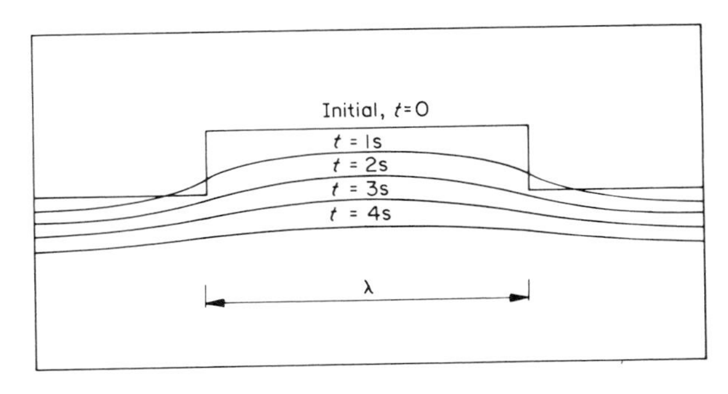

Fig. 6.6 Successive anode profiles during smoothing of initially rectangular step (after Fitz-Gerald and McGeough [3])

will be shown to dissolve more quickly than those which are less sharp, and widely spaced irregularities will be dissolved more rapidly than those close together (the basic shape being the same in all cases). Numerical illustrations of such considerations are now given.

Examples

Typical values used for the various constants are: $V = 10$ V, $p = 0{\cdot}5$ mm, $f = 5{\cdot}67 \times 10^{-3}$ mm/s, $M = 1{\cdot}7 \times 10^{-5}$ cm^2 V^{-1} s^{-1}.

(i) Consider first the smoothing of cusp-shaped irregularities described by

$$y_a = h - \frac{\epsilon(0)x^4}{\lambda^4} \quad (-\lambda \leqslant x \leqslant \lambda)$$

with the same notation as before. We use a cosine expansion:

$$y_a = p(t) + \epsilon(0) \sum_1^\infty a_n(t) \cos \frac{n\pi x}{\lambda}$$

where

$$a_n(0) = \frac{8(-1)^n}{(n\pi)^4}[(n\pi)^2 - 6] \quad (n \neq 0)$$

For $2\lambda = 10^{-1}$ mm, the time dependence of ϵ is shown in Fig. 6.7. For comparison, the behaviour of a sinusoidal irregularity of the

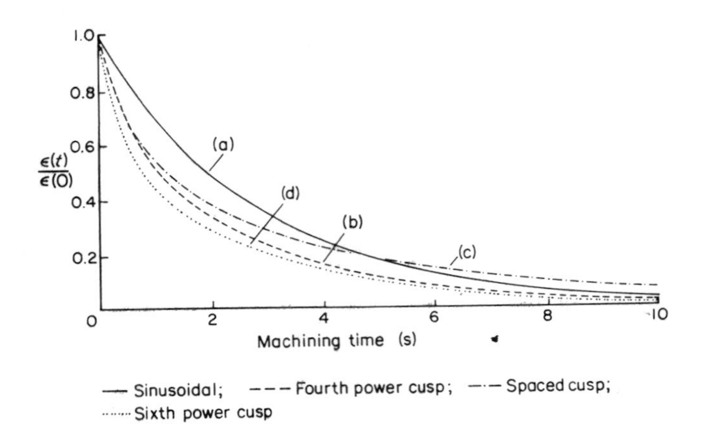

Fig. 6.7 Decrease in height of irregularities with time, showing dependence of smoothing times on shape and spacing: (a) sinusoidal; (b) fourth power cusp; (c) spaced cusp; (d) sixth power cusp (after Fitz-Gerald and McGeough [3])

same wavelength is also given. Table 6.1 shows the initial odd coefficients (which determine the height of the irregularities).

Table 6.1 Initial odd Fourier coefficients for cusp-shaped irregularity. [Section 6.10. Example (i)]

$a_1 = 0{\cdot}316$	$a_9 = 0{\cdot}010$
$a_3 = 0{\cdot}084$	$a_{11} = 0{\cdot}007$
$a_5 = 0{\cdot}031$	$a_{13} = 0{\cdot}005$
$a_7 = 0{\cdot}016$	$a_{15} = 0{\cdot}004$

(ii) Separation of the cusps by plane segments of length 2λ is next considered. The basic wavelength is now 0·2 mm. The relevant Fourier coefficients are given in Table 6.2.

Table 6.2 Initial Fourier coefficients. [Section 6.10. Example (ii)]

$a_1 = 0{\cdot}188$	$a_9 = 0{\cdot}019$
$a_3 = 0{\cdot}121$	$a_{11} = 0{\cdot}013$
$a_5 = 0{\cdot}058$	$a_{13} = 0{\cdot}009$
$a_7 = 0{\cdot}031$	$a_{15} = 0{\cdot}007$

Note that the higher harmonics are relatively more important in this case. Initially, smoothing will be faster since there is a greater field concentration. Later, as the shape approaches the basic sinusoidal form, the rate of smoothing will decrease; since the basic wavelength is double that of the unseparated cusps, the rate will be less than the rate for the unseparated cusps (Fig. 6.7). It will be noted that, for this wavelength, the initially greater smoothing rate persists for only half a second. This effect is more pronounced for longer wavelengths, and Fig. 6.8 shows the comparison for a basic wavelength of 40 mm (an extreme case).

(iii) The effect of shape, as distinct from spacing, on the smoothing rate for the cusps is now considered. The results obtained from example (i) will be compared with those for a sharper cusp, whose initial shape is described by:

$$y_a = p(0) - \epsilon(0)\frac{x^6}{\lambda^6} \quad (-\lambda \leqslant x \leqslant \lambda)$$

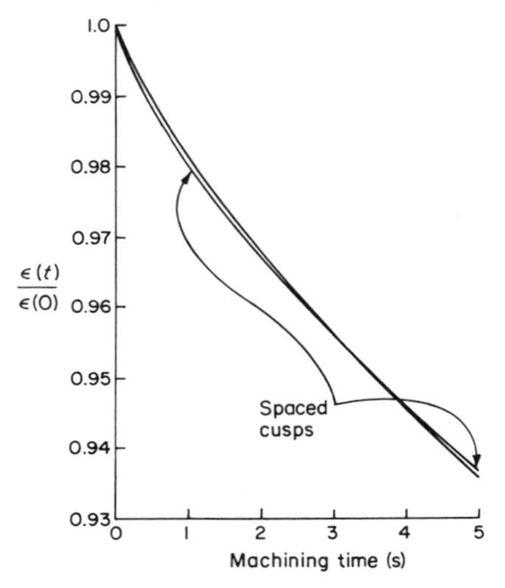

Fig. 6.8 Dependence of spacing for fourth power cusps of relatively long wavelength ($2\lambda = 40$ mm) (after Fitz-Gerald and McGeough [3])

again with $2\lambda = 10^{-1}$ mm. The Fourier coefficients are given in Table 6.3. The higher harmonics are again of greater importance; smoothing will again be faster than in (i) (Fig. 6.7).

Table 6.3 Initial Fourier coefficients. [Section 6.10. Example (iii)]

$a_1 = 0{\cdot}250$	$a_9 = 0{\cdot}015$
$a_3 = 0{\cdot}107$	$a_{11} = 0{\cdot}010$
$a_5 = 0{\cdot}045$	$a_{13} = 0{\cdot}007$
$a_7 = 0{\cdot}024$	$a_{15} = 0{\cdot}005$

(iv) Finally, the smoothing of a sinusoidal and a square-wave rectangular irregularity of the same wavelength is considered (Fig. 6.9). Since there is less field concentration at the middle of the square wave, its smoothing rate will be smaller. The behaviour of a

Table 6.4 Fourier coefficients. [Section 6.10. Example (iv)]

Equally spaced		Separated	
$a_1 = 0{\cdot}637$	$a_9 = 0{\cdot}071$	$a_1 = 0{\cdot}551$	$a_9 = 0{\cdot}000$
$a_3 = -0{\cdot}212$	$a_{11} = -0{\cdot}058$	$a_3 = 0{\cdot}000$	$a_{11} = -0{\cdot}050$
$a_5 = 0{\cdot}127$	$a_{13} = 0{\cdot}049$	$a_5 = -0{\cdot}110$	$a_{13} = 0{\cdot}042$
$a_7 = -0{\cdot}091$	$a_{15} = -0{\cdot}042$	$a_7 = 0{\cdot}079$	$a_{15} = 0{\cdot}000$

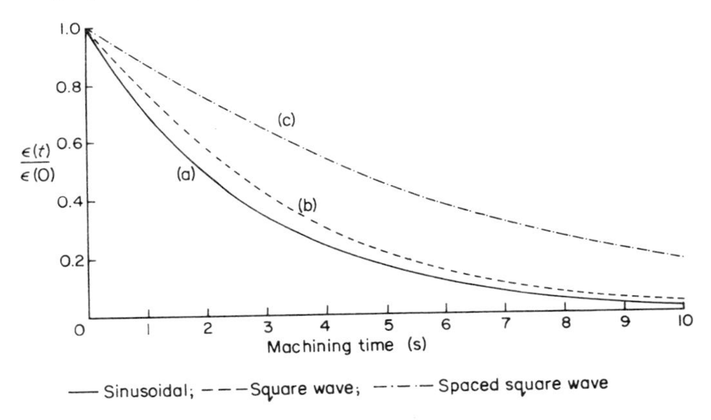

Fig. 6.9 Smoothing of rectangular steps, showing dependence on spacing; the curve for a sinusoidal wave is shown for comparison; (a) sinusoidal, (b) square wave, (c) spaced square wave (after Fitz-Gerald and McGeough [3])

spaced square wave is also given; in this case, no indication of the initially greater smoothing rate found for the cusp example is seen, and even for a wavelength of 1 mm the smoothing rate is always smaller. Table 6.4 shows a comparison between the Fourier coefficients for equally spaced, and separated, square waves.

6.11 The effects of overpotentials

6.11.1 Overpotentials only at the cathode

For clarity, the effect of an overpotential only at the cathode is initially considered. From Section 6.2, the boundary condition for the potential at the cathode is now given as $\phi = f(J)$. Expanding $f(J)$ as a Taylor series, we obtain

$$\begin{aligned}\phi &= f(J) + \left.\frac{\partial f}{\partial J}\right|_{J=\bar{J}} (J - \bar{J}) \\ &= \alpha + \beta(J - \bar{J}) \qquad (6.53)\end{aligned}$$

where $\bar{J}$ is the average current density, α (the average overpotential) is $f(\bar{J})$, and

$$\beta = \left.\frac{\partial f}{\partial J}\right|_{J=\bar{J}} \qquad (6.54)$$

The small difference between the current density, J, and the average current density is a consequence of the overpotential.

Because of the overpotential, the previous expression for the potential, Equation (6.22), must be modified. It is possible to consider the effect to be 'reflected' from the anode; the shape variations on the anode cause variations in the field concentrations there, and these in turn create small variations in the current density at the cathode. It is therefore plausible to consider a new potential of the form

$$\phi = A + \frac{V - A}{p}$$

$$\times \left\{ y - \epsilon(0) \sum_1^\infty \sin \frac{n\pi x}{\lambda} \frac{a_n \sinh(n\pi y/\lambda) + b_n \sinh[n\pi(p - y)/\lambda]}{\sinh(n\pi p/\lambda)} \right\} \quad (6.55)$$

where A and b_n are to be determined.

An analysis of the problem has been set out by Fitz-Gerald and McGeough [3]. Again, for brevity, only the salient results are given here.

Owing to overpotential, the usual equation for the variation in gap width becomes modified:

$$\frac{dp}{dt} = M\left(\frac{V - A}{p}\right) - f \quad (6.56)$$

in which it can be shown that $A = \alpha$, the average overpotential.

From Equation (6.56), the condition for the equilibrium gap is now

$$f = \frac{MV}{p} - \frac{MA}{p} \quad (6.57)$$

Clearly, the feed-rate required for a specified gap is now reduced by an amount MA/p. A slower anode dissolution rate then occurs, and consequently a slower overall reduction rate of the height of the irregularities.

Example

Suppose $f(J) = bJ$ (low current density form of Tafel's relationship). Then, in Equation (6.53) $\alpha = b\bar{J}$ and $\beta = b$; and now Equations (6.53) and (6.56) yield

$$\frac{dp}{dt} = \frac{MV}{p + b\kappa_e} - f$$

i.e. the effect is to increase the effective gap width by a factor $(1 + b\kappa_e/p)$. Alternatively, the time required to achieve the equilibrium gap will be increased. The significance in electrodeposition of the quantity b has been fully recognised by, for example, Kasper [6–8] and Hoar and Agar [9].

The presence of overpotentials also causes an alteration in the Fourier coefficients which are used to describe the anode surface. For constant p and for odd-shaped irregularities, these coefficients are altered to

$$a_n(t) = a_n(0) \exp\left[-\frac{M(V-A)}{p}\frac{n\pi}{\lambda}\coth\frac{n\pi p}{\lambda}\,\omega(p)t\right] \quad (6.58)$$

where $\omega(p)$ is a correction factor for the overpotential.

Fitz-Gerald and McGeough discuss two particular conditions for $\omega(p)$. When the irregularities are of long wavelength,

$$\omega = \left(1 + \frac{\mu}{\sigma}\right)^{-1} \quad (6.59)$$

For convenience, two dimensionless parameters have been introduced:

$\sigma = p/\lambda$ the configuration parameter

$\mu = \beta\kappa_e/\lambda$ the overpotential parameter

For $\mu/\sigma \ll 1$, $\omega \simeq (1 - \mu/\sigma)$, which introduces a small correction to the smoothing time. For $\mu/\sigma \gg 1$, however, as here, long-wavelength irregularities (or components) are dissolved very slowly; smoothing is then difficult. (Note: a corresponding situation in electrodeposition has been discussed fully by Wagner [10].)

In the short-wavelength case, ω becomes

$$\omega = \frac{1 + n\pi\mu(1 - 4\,e^{-2n\pi\sigma})}{1 + n\pi\mu} \quad (6.60)$$

If $n\pi\mu \ll 1$, this gives $\omega = 1 - 4n\pi\mu\,e^{-2n\pi\sigma}$, while for $n\pi\mu \gg 1$, $\omega = 1 - 4\,e^{-2n\pi\sigma}$. In both cases, ω is negligibly different from unity, in view of the short-wavelength assumption. These short-wavelength irregularities are therefore unaffected by the fluctuations in the overpotential, although the time constant for their rate of removal is increased by a factor $V/(V - A)$.

The overall effect of a current density-dependent overpotential, then, is to produce waves of relatively long wavelength on the smoothed surface. The persistence of these waves depends on the value of μ/σ, as demonstrated in the long-wavelength case above. The effect may be diminished by either (a) reducing κ_e or (b) increasing p (or decreasing the feed-rate f). Both of these remedies have the unwanted side-effect of reducing the current, so that elimination of overpotential effects may only be obtained at the expense of increased smoothing time.

6.11.2 *Overpotentials at both electrodes*

If now overpotentials on both anode and cathode are to be considered, the boundary conditions become

$$\phi = f(J) \qquad \text{on the cathode}$$

$$\phi = V - g(J) \qquad \text{on the anode}$$

As before, $f(J)$ and $g(J)$ are expanded as Taylor series, and all but the first two terms of each are neglected.

$$f(J) = \alpha + \beta(J - \bar{J}) \tag{6.61}$$

$$g(J) = \gamma + \tau(J - \bar{J}) \tag{6.62}$$

where α and β are defined as before, and $\gamma = g(\bar{J})$, $\tau = \partial g/\partial J\,|_{J=\bar{J}}$.

By arguments similar to those already outlined, it can be postulated that effects of these overpotentials can be described in terms of two additional Fourier coefficients b_n and c_n, and overpotential parameters, μ and σ, defined above, and $\nu = \tau\kappa_e/\lambda$.

6.11.3 *Overpotentials only at the anode*

The condition $\nu \ll \mu$ corresponds to overpotentials only at the cathode. Their effects have been discussed. However, if $\nu \gg \mu$, it can be shown that in the limiting case, $\mu/\nu \to 0$,

$$b_n = 0$$

$$c_n = -a_n \frac{n\pi\nu}{\tanh n\pi\sigma + n\pi\nu} \tag{6.63}$$

The case now is that of overpotential only at the anode. For long wavelengths,

$$c_n = -a_n \frac{\nu}{\sigma + \nu} \tag{6.64}$$

and the behaviour is similar to that in the corresponding case discussed above.

The short-wavelength case, however, gives

$$c_n = -a_n \frac{n\pi\nu}{1 + n\pi\nu} \tag{6.65}$$

Whereas previously the overpotential had little or no effect on the short-wavelength irregularities (other than the overall rate reduction), the large overpotential case ($\nu \gg 1$) gives

$$c_n \simeq -a_n$$

yielding

$$\frac{da_n}{dt} \simeq 0 \tag{6.66}$$

for all n. Thus the reduction rate for all Fourier components is very small; the anode profile is 'frozen' as the surface is machined away. Even in the case $\nu \ll 1$, when the effect on the long wavelengths decreases, for suitably large n, $da_n/dt \to 0$; in this situation, the short-wavelength components are preferred to the long-wavelength ones.

An experimental verification of this theory remains to be published. The results, however, do demonstrate the type of analysis required for the shaping problem in ECM. In addition, the investigation of overpotentials indicates the effects which might be encountered in ECM, particularly since close resemblance is obtained between these results and those already found theoretically and experimentally in electrodeposition.

References

1. McGeough, J. A. and Rasmussen, H., *J. Inst. Maths. and Applics.* (1974) **1**, 13.
2. Rasmussen, H. (in preparation).
3. Fitz-Gerald, J. M. and McGeough, J. A., *J. Inst. Maths and Applics.* (1969) **5**, 389.
4. Warburton, P. and Sadler, B., Paper presented at First Int. Conf. on ECM, Leicester University, Mar. 1973.
5. Tipton, H., Proc. Fifth Int. Conf. Mach. Tool Res., Birmingham (1964) Sept., p. 509.
6. Kasper, C., *Trans. Electrochem. Soc.* (1940) **77**, 353, 365.
7. Kasper, C., *Trans. Electrochem. Soc.* (1940) **78**, 131, 147.
8. Kasper, C., *Trans. Electrochem. Soc.* (1942) **82**, 153.
9. Hoar, T. P. and Agar, J. N., *Discuss. Faraday Soc.* (1947) **1**, 162.
10. Wagner, C., *J. Electrochem. Soc.* (1951) **98**, 116.

CHAPTER SEVEN

Anodic Shaping and Cathode Shape Design

The two fundamental problems in ECM are (i) the prediction of the resultant anode shape when the cathode shape is known, and (ii) the design of the cathode shape to achieve a required anode shape. The second problem is by far the more difficult, and it can be regarded as the primary one of ECM.

Basic approaches for tackling the two problems are of interest. First, analytic solutions are possible, although they have proved to be limited in their applicability. Because of this drawback, the 'cos θ' method for cathode design has been developed; in addition, analogue techniques have been tried, although they have their own limitations. The best means of solution to the problems of shaping in ECM seems to be offered by numerical methods. Even these, however, seldom incorporate effects often encountered in ECM, for instance, those arising from the electrolyte flow. The inclusion of such phenomena in any analysis would render almost intractable the general problem of electrochemical shaping. Indeed, the lack of progress in solving this problem has meant that empirical methods are still widely used in practice.

7.1 Solutions by analysis

As in the previous chapter, three basic equations are used.

(i) Laplace's equation:

$$\nabla^2 \phi = 0 \tag{7.1}$$

the solution of which gives the potential ϕ in the electrolyte, particularly at the electrode surfaces.

(ii) Ohm's law:

$$J = -\kappa_e \nabla \phi \qquad (7.2)$$

where J, the current density, will be found from the potential from Equation (7.1), κ_e being the electrolyte conductivity.

(iii) Faraday's law:

$$\dot{r}_a = \left(\frac{A}{z\rho_a F}\right) J \qquad (7.3)$$

in the usual notation, which is used to find the anodic dissolution rate.

Two procedures are known to be available for solving these equations.

7.1.1 A perturbation method [1]

This method has been used to describe electrochemical shaping when the amplitude of shapes on the cathode and anode is small compared with the inter-electrode gap. The small size of the shapes on the electrodes will be shown below to restrict the use of this method to a limited range of shaping problems. Nevertheless, this treatment usefully demonstrates characteristic behaviour encountered in electrochemical shaping; and, moreover, it can be applied to any electrode shape. It is also one of the few treatments which allows the role of overpotentials to be examined. This method obviously has a close resemblance to the main analysis of anodic smoothing in Chapter 6.

In the previous chapter, the anode irregularities were assumed to be periodic, and a Fourier series expansion was used to describe their behaviour. In this chapter, a more general class of electrode shapes (and, in particular, isolated ones) will be taken, and the corresponding analysis will be performed by means of Fourier transforms.

Consider, first, a case where the cathode shape is known, and the resultant shape on the anode is to be found. Suppose the cathode shape is given by the Fourier integral

$$y_c = \epsilon(0) \int_{-\infty}^{\infty} b(k) e^{ikx} \, dk \qquad (7.4)$$

which represents a small deviation from $y_c = 0$; i.e. $|y_c|/p \ll 1$, where p is the average inter-electrode gap, and $\epsilon(0)$ is the size of irregularity on the electrode surface. Suppose further that the anode is initially plane. (This is not essential to the argument, but simplifies the algebra). After a time t, the anode shape may be described by

$$y_a = p + \epsilon(0) \int_{-\infty}^{\infty} a(k,t) e^{ikx} \, dk \tag{7.5}$$

The shapes on both anode and cathode will influence the inter-electrode potential which, it is assumed, may be written as a first-order perturbation to the potential between two plane parallel electrodes:

$$\phi = \phi_0 + \frac{\epsilon(0)}{p} \phi_1$$

where ϕ_0 satisfies the boundary conditions $\phi_0 = 0$ on $y = 0$, $\phi_0 = V$ on $y = p$, and ϕ_1 satisfies

$$\begin{aligned} \phi_0 + \frac{\epsilon(0)}{p} \phi_1 &= 0 \quad \text{on } y = y_c \\ \phi_0 + \frac{\epsilon(0)}{p} \phi_1 &= V \quad \text{on } y = y_a \end{aligned} \tag{7.6}$$

By procedures similar to those in Chapter 6, an expression can be obtained for the potential:

$$\phi = \frac{Vy}{p} - V \frac{\epsilon(0)}{p} \int_{-\infty}^{\infty} e^{ikx} \frac{a \sinh ky + b \sinh k(p-y)}{\sinh kp} \, dk$$

for the rate of change of gap,

$$\frac{dp}{dt} = \frac{MV}{p} - f$$

and for the time-dependence of the Fourier coefficients,

$$\frac{\partial a}{\partial t} = -\frac{MVk}{p}(a \coth kp - b \operatorname{cosech} kp)$$

For constant p, the above equation can be solved to show that the amplitude of any frequency component on the anode tends to the limiting value

$$a(k) = b(k) \operatorname{sech} kp \tag{7.7}$$

as $t \to \infty$. For $a(k,0) = 0$ (initially smooth anode), it is also pointed out that components with large $|k|$ will have little effect on the profile produced on the anode. The range of possible anode shapes which may be predicted by this method of analysis is therefore restricted. A fuller discussion of these limitations is given elsewhere [1].

When this theory is extended to cover cathode design, further stringent restrictions are imposed. Indeed, the usefulness of this theory for studies of cathode design becomes exceedingly limited. The theory, however, does indicate the approaches required for analytic solutions of the shaping problem in ECM. Moreover, it emphasises the difficulty of the problem.

One further result from that work is of interest: the analysis can be extended to include the effects of overpotential. For an arbitrary current density-dependent overpotential at the cathode only, the limiting amplitude on the anode is shown to be modified to

$$a(k) = b(k) \frac{\operatorname{cosech} kp}{\coth kp + (\mu/\sigma)kp} \tag{7.8}$$

where μ/σ ($= \beta\kappa_e/p$), the dimensionless parameter for the cathode overpotential, previously introduced in Chapter 6, has again been used. Clearly, the limiting amplitude has been multiplied by a factor $\coth kp/[\coth kp + (\mu/\sigma)kp]$, and the profile machined on the anode becomes 'blurred'.

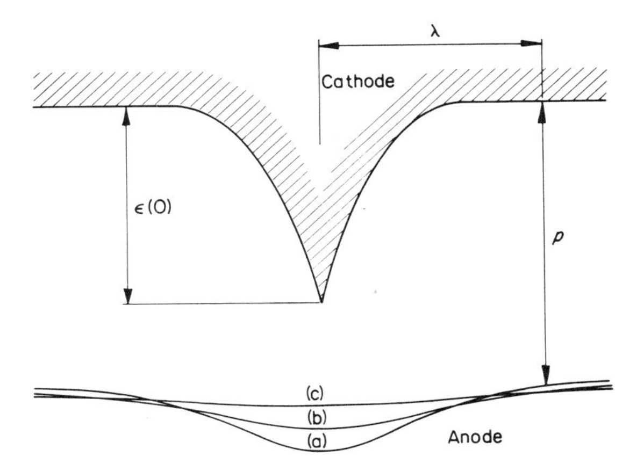

Fig. 7.1 Shape produced on anode by cusp shaped cathode. Shape (a), without overpotential; shapes (b) and (c), with cathode overpotential; $\lambda = 1$ mm; $p = 0{\cdot}5$ mm; vertical scale magnified twenty times (after Fitz-Gerald et al. [1])

Overpotentials only at the anode can be shown to have no effect on the limiting amplitude on the anode shape. But overpotentials at either electrode are shown to increase the time of ECM necessary to achieve the limiting amplitude on the anode.

Example

Figure 7.1 indicates the shape produced on the anode by an isolated cusp on the cathode. Curve (a) represents the anode profile, when overpotentials are not considered. Curves (b) and (c) demonstrate the 'blurring' of this profile by overpotentials at the cathode only. The amount of this overpotential is defined by the quantity μ/σ $(= \beta\kappa_e/p)$: $\mu/\sigma = 1$ [curve (b)]; $\mu/\sigma = 10$ [curve (c)].

7.1.2 A complex variable method

Methods of solutions of Laplace's equation with specified boundary conditions by this method are well known, and can be found in any of the standard textbooks. In brief, the basis of any such approach is as follows.

Suppose a solution is sought for the two-dimensional Laplace's equation:

$$\frac{\partial^2 \phi}{\partial x^2} + \frac{\partial^2 \phi}{\partial y^2} = 0$$

To solve this equation, we put $z = x + iy$, where $i = \sqrt{-1}$. Next, consider any function of z, which is written

$$\omega = f(z) = f(x + iy) = \phi(x, y) + i\psi(x, y) \tag{7.9}$$

That is, ϕ and ψ are real and imaginary parts of $f(z)$.

By partial differentiation of $\omega = f(x + iy)$:

$$\frac{\partial^2 \omega}{\partial x^2} + \frac{\partial^2 \omega}{\partial y^2} = 0 \tag{7.10}$$

Separation of the real and imaginary parts yields

$$\nabla^2 \phi = 0 \quad \text{and} \quad \nabla^2 \psi = 0 \tag{7.11}$$

Hence ϕ and ψ are possible potential functions satisfying Laplace's equation. Note that, on partial differentiation of Equation (7.9),

$$\frac{\partial \omega}{\partial y} = if'(z) = i\frac{\partial \omega}{\partial x}$$

That is,

$$\frac{\partial \phi}{\partial x} = \frac{\partial \psi}{\partial y} \quad = \text{real part of } f'(z) \tag{7.12}$$

$$\frac{\partial \psi}{\partial x} = -\frac{\partial \phi}{\partial y} = \text{imaginary part of } f'(z) \tag{7.13}$$

Solutions to Laplace's equation can now be obtained either directly from known functions, e.g. $f(z) = z^3$, the boundary conditions being known, so that equipotential lines can be defined, or by means of the method of electrical images, or by conformal mapping. The last mentioned approach is most effective for ECM problems. Its principle consists of the transformation of one problem into another which is tractable.

Thus, the transformation $\zeta = f(z)$ transforms a point z in the (x,y)-plane into a point $\zeta = \xi + i\eta$ in the (ξ,η)-plane. Since $\mathrm{d}\zeta = f'(z)\mathrm{d}z$, $|\mathrm{d}\zeta| = |f'(z)|\,|\mathrm{d}z|$, $\arg \mathrm{d}\zeta = \arg f'(z) + \arg \mathrm{d}z$.

Two points can now be made: (i) in the vicinity of the point $z = z_0$, where $\zeta = \zeta_0 = f(z_0)$, all distances in the ζ-plane are $|f'(z_0)|$ times greater than the corresponding distances in the z-plane; (ii) small arcs are turned through an angle $\arg f'(z_0)$. Thus a small element of area in the z-plane becomes an element in the ζ-plane. The latter element has the same shape, but has dimensions magnified by $|f'(z_0)|$, the element being rotated through an angle $\arg f'(z_0)$.

Thus, if a function $\phi(x,y)$ can be expressed in terms of ξ and η, $\Phi(\xi,\eta)$, by the transformation $\zeta = f(z)$ then

$$\frac{\partial^2 \Phi}{\partial \xi^2} + \frac{\partial^2 \Phi}{\partial \eta^2} = \frac{1}{\left|\frac{\mathrm{d}\zeta}{\mathrm{d}z}\right|^2}\left(\frac{\partial^2 \phi}{\partial x^2} + \frac{\partial^2 \phi}{\partial y^2}\right) \tag{7.14}$$

Suppose

$$\frac{\partial^2 \phi}{\partial x^2} + \frac{\partial^2 \phi}{\partial y^2} = 0$$

Then

$$\frac{\partial^2 \Phi}{\partial \xi^2} + \frac{\partial^2 \Phi}{\partial \eta^2} = 0$$

so that Φ is a potential function in the (ξ,η)-plane.

In the application of complex variable techniques to the shaping problem in ECM, the three equations (7.1), (7.2), and (7.3) are again used. As usual, any analysis is eased by the assumption of steady-state machining conditions, and by the exclusion of the effects of electrolyte flow and overpotentials.

For these conditions, the potential boundary conditions are given by Equations (6.4) and (6.5):

$$\phi = 0 \quad \text{on the cathode}$$

$$\phi = V \quad \text{on the anode}$$

Since steady-state ECM is taking place, a further condition on the anode is that the normal dissolution rate of the anode surface equals the cathode feed-rate in that direction. That is,

$$\kappa_e \left(\frac{\partial \phi}{\partial n} \right) = f \cos \theta \tag{7.15}$$

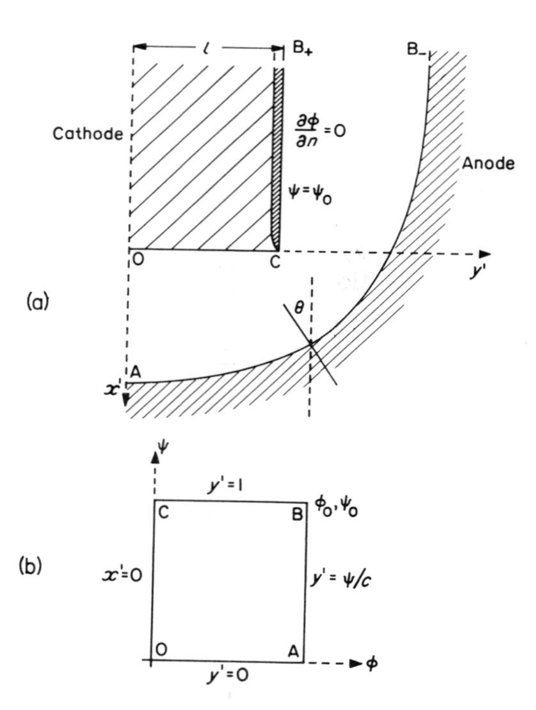

Fig. 7.2 (a) Configuration for a plane-faced cathode tool with insulation; (b) boundary conditions in the ω-plane (after Collett et al. [2])

where $(\partial\phi/\partial n)$ is the normal component of the electric field at the anode surface, f is the cathode feed-rate (in the vertical direction), and θ is the angle between the normal to the anode boundary and the direction of cathode movement.

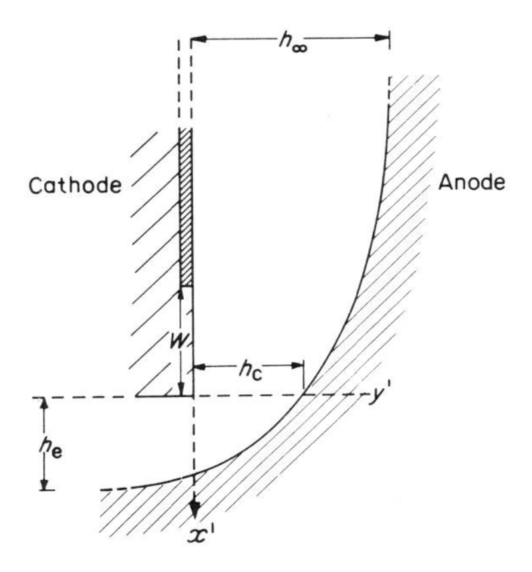

Fig. 7.3 Straight-sided cathode with finite uninsulated land width at base and insulated upper length; $W = 1{\cdot}86h_e$; $h_\infty = 3{\cdot}36h_e$ (after Hewson-Browne [3])

These approaches have been used to find the equilibrium anode shapes resulting from two particular cathode shapes: (i) a plane-faced cathode with complete insulation on its side walls (Fig. 7.2), and (ii) a partly uninsulated, straight-sided cathode (Fig. 7.3) [2, 3].

From the methods of complex variables, the electric field in the electrolyte between the electrodes is

$$E = (E_x, E_y)$$

and

$$\bar{E} = E_x - iE_y = \frac{d\omega(z)}{dz}$$

where $z = x + iy$, and $\omega = \phi + i\psi$.

The boundary conditions (6.4) and (6.5) and (7.15) apply. On the insulated surfaces of the cathode, the normal component of the electric field is zero, so that

$$\psi = \text{constant} \tag{7.16}$$

on these surfaces.

Another form of Equation (7.15) is useful:

$$\frac{\partial \phi}{\partial n} = \tau \cos \theta \tag{7.17}$$

where τ is a constant, obtained from the machining parameters.

If s is the arc length measured along the anode boundary S in the direction of increasing y',

$$\cos \theta = \frac{\partial y'}{\partial s}$$

and

$$\frac{\partial \phi}{\partial n} = \frac{\partial \psi}{\partial s}$$

Thus, on integration the boundary condition (7.17) becomes

$$\psi = \tau y' \tag{7.18}$$

Since the electrode configuration [Fig. 7.2(a)] is symmetrical about the x'-axis, and since, from Equation (7.18), $\psi = 0$ at A, we have that $\psi = 0$ on OA.*

Also, from Equation (7.16),

$$\psi = \psi_0 \text{ on } B_+C$$

Thus, only the region OABCD needs consideration, and that region is now mapped onto the ω-plane. The boundary conditions in the ω-plane are shown in Fig. 7.2(b).

A solution for ω of the form

$$z = \frac{\omega}{\tau} + \sum_{n=0}^{\infty} a_n \sin\left[(2n+1)\frac{\pi\omega}{2V}\right] \tag{7.19}$$

is now sought so that conditions on OA, AB, and CO are satisfied. Here a_n are real constants.

When this analysis is carried out, the configuration of electrodes can be shown to be specified by the parameters V, τ, and l, where $2l$ is the width of the cathode. When some typical values, $V/\tau = 1{\cdot}014l$, $h_e = 0{\cdot}899l$, and $h_\infty = 1{\cdot}006l$, where h_∞ is the total overcut at B_+B_-, are used to describe the limiting anode surface, one further result of

* In the study on which this treatment is based [2] the x and y axes are interchanged from the convention adopted elsewhere in this chapter. To maintain the sequence of argument, but at the same time to avoid confusion, the symbols x' and y' are used here; so that elsewhere $x = y'$ and $y = x'$.

interest can be obtained. It predicts that the overcut at the corner C of the cathode is related to the equilibrium gap by the relation

$$\frac{\text{overcut at corner C}}{\text{equilibrium gap (OA)}} \simeq 0{\cdot}731 \tag{7.20}$$

For a cathode without insulation along the side BC, further analysis by mapping yields that

$$\frac{\text{overcut at corner C}}{\text{equilibrium gap (OA)}} = 1{\cdot}159 \tag{7.21}$$

This method of analysis has been extended to cover a wider class of cathode tools in which the straight-sided cathode has a finite un-insulated land width W at its base as well as an insulated portion [3] (see Fig. 7.3).

Two consecutive mappings are used to show that the variation of the total overcut h_∞ with equilibrium gap is given by

$$\frac{h_\infty}{h_e} = 1 + \frac{1}{4}\left(\frac{3\pi W}{h_e}\right)^{2/3} + \cdots \tag{7.22}$$

for $W \ll h_e$, and

$$\frac{h_\infty}{h_e} \simeq \left(\frac{2W}{h_e}\right)^{1/2} + \frac{2}{\pi}\log 2 + \cdots \tag{7.23}$$

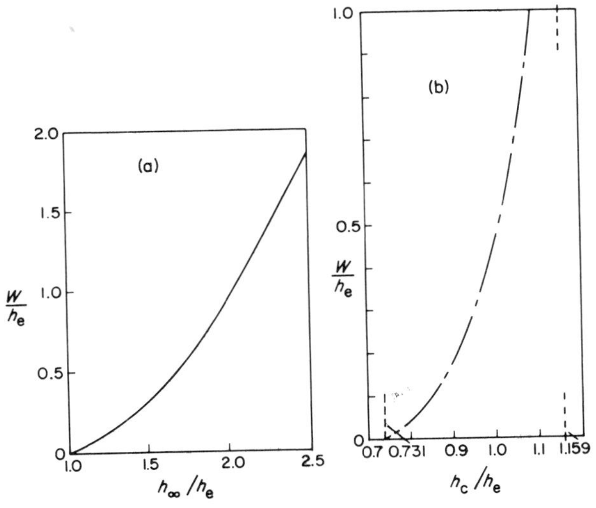

Fig. 7.4 Land width W as a function of (a) total overcut h_∞, and (b) overcut at the corner of the cathode, h_c (after Hewson-Browne [3])

for $W \gg h_e$. The variation in the total overcut h_∞ with the land width W is shown in Fig. 7.4(a), whilst the dependence of the overcut at the corner on land width is given in Fig. 7.4(b).

In practice, the overcut in electrochemical hole-drilling can be reduced, and the accuracy of drilling improved, by the use of a passivating electrolyte, which yields current efficiencies which are low and high, respectively, at low and high current densities. The former condition is achieved in the region of the side gap, whilst high current efficiencies at the high current densities are obtained along the main, front machining gap.

7.2 Solution by the 'cos θ' method

The limited availability and usefulness of analytic solutions to the shaping problem have meant a search for alternative methods which are simpler to handle and which are useful in tackling a range of practical problems. Accordingly, the so-called 'cos θ' method has received considerable attention [4–6]. Again, this procedure is based on steady-state machining conditions, and it excludes consideration of electrolyte flow and overpotential effects. Its main features can be explained with reference to Fig. 7.5.

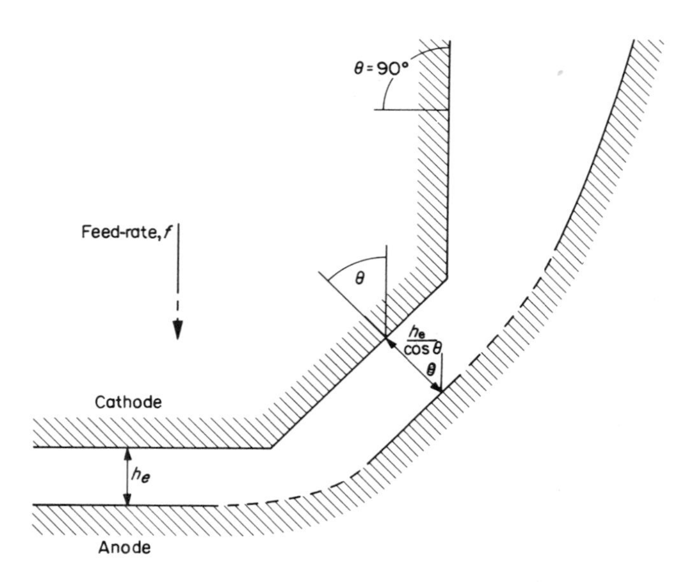

Fig. 7.5 Principle of the 'cos θ' method (after Tipton [4])

Consider a cathode tool which consists of several planar sections inclined at different angles, θ. The equilibrium gap between any section of the cathode surface and that corresponding surface of the anode which is parallel to it is $h_e/\cos\theta$. As before, θ is measured between the normal to the anode surface and the direction of cathode feed. The 'cos θ' method of cathode design can be applied to electrode regions where the electric field can be assumed to be normal to the surfaces of the electrodes. It cannot, of course, be applied to regions corresponding to discontinuities at the cathode surface, and where the angle θ is great ($\simeq 90°$); that is, its application is limited to regions where the local radii of curvature of the anode and cathode surfaces are large compared with the equilibrium gap.

The following parametric equations have proved useful for calculating the cathode shape necessary for a required anode shape [4].

Suppose that the anode shape is given by

$$y = f(x) \tag{7.24}$$

Any point, $A(x, y)$, say, on the anode surface then corresponds to an

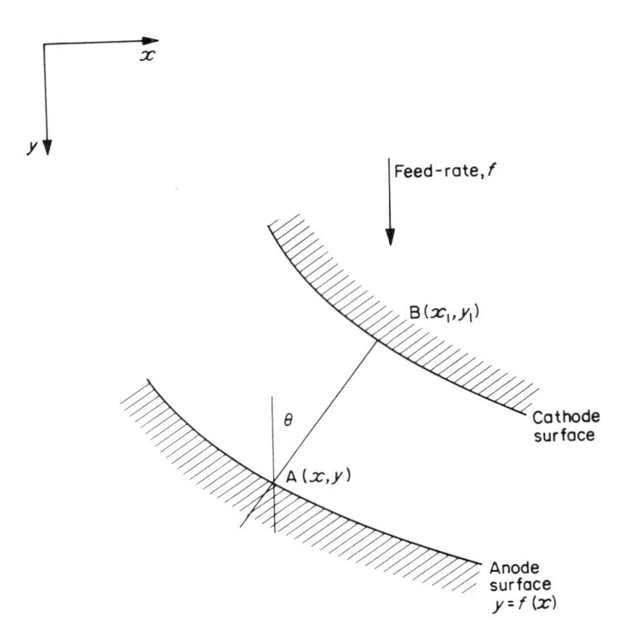

Fig. 7.6 Cathode design by 'cos θ' method

equivalent point $B(x_1,y_1)$ on the cathode, the gap width between A and B being $h_e/\cos\theta$ (Fig. 7.6). We have

$$\begin{aligned} y - y_1 &= \mathrm{AB}\cos\theta \\ &= h_e \end{aligned} \tag{7.25}$$

and

$$\begin{aligned} x_1 - x &= \mathrm{AB}\sin\theta \\ &= h_e\tan\theta \\ &= h_e\left(\frac{\mathrm{d}y}{\mathrm{d}x}\right) \end{aligned} \tag{7.26}$$

Suppose, now, that a general expression for the anode surface is

$$y = a + bx + cx^2 \tag{7.27}$$

Use of the above equations then yields

$$y_1 + h_e = a + bx_1 + cx_1^2 - bh_e\frac{\mathrm{d}y}{\mathrm{d}x} - 2cx_1h_e\frac{\mathrm{d}y}{\mathrm{d}x} + ch_e^2\left(\frac{\mathrm{d}y}{\mathrm{d}x}\right)^2 \tag{7.28}$$

Now from Equations (7.26) and (7.27),

$$\frac{\mathrm{d}y}{\mathrm{d}x} = \frac{b + 2cx_1}{1 + 2ch_e} \tag{7.29}$$

and substitution of Equation (7.29) into (7.28) gives

$$y_1 = a + bx_1 + cx_1^2 - h_e - h_e\left[\frac{(b + 2cx_1)^2}{1 + 2ch_e}\right] \tag{7.30}$$

in which terms of order $h_e^2(\mathrm{d}y/\mathrm{d}x)^2$ have been assumed negligible.

In Equation (7.30) the first four terms still describe the anode surface, but indicate its displacement through a distance h_e. The final term includes the correction for the curvature of the anode.

The graphical and analytic means of predicting electrode shapes by the 'cos θ' method have largely been replaced by computational techniques.

7.3 Analogue methods of solution

Analogue methods for solving Laplace's equation are well known, and their application to the ECM problem of anode shaping has been the subject of several investigations. In all these studies, the same

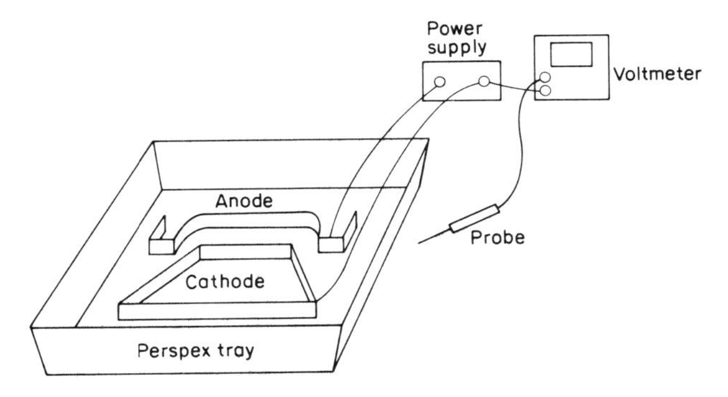

Fig. 7.7 Equipment for electrolytic tank analogue

principles are used. One electrode boundary shape is fixed and known (for example, the cathode). The potential conditions between the electrodes and at their boundaries are also known. The other electrode shape is now sought so that certain boundary conditions are met. These boundary conditions, as usual, are

$$\phi = 0 \quad \text{on the cathode}$$

$$\phi = V \quad \text{constant on the anode}$$

$$\frac{\partial \phi}{\partial n} = \text{constant at the anode}$$

7.3.1 *'Electrolytic tank' analogue method*

The basic equipment consists of a shallow perspex tray, filled with electrolyte (e.g. copper sulphate solution), and thin metal strips (e.g. copper) which can be clamped in position in the tray. The strips represent the anode and cathode boundaries. Insulated boundaries are likewise represented by perspex strips. A small voltage (1 to 3 V) is applied between the electrodes from a stabilised power supply. It is measured with a high input impedance voltmeter connected to a probe, so prepared that potential differences between the head of the probe and the electrode boundary are reduced as much as possible (see Fig. 7.7).

Suppose, for example, that the cathode shape is fixed and the anode shape is to be found. Enlarged scale drawings of the shape of the known boundary and the *probable* equilibrium shape of the anode are then prepared, preferably on graph paper. On the anode boundary a control point is fixed from which a line normal to the known cathode boundary can be drawn. At points along the anode boundary, the ratios are noted of the normal potential gradients relative to the fixed control point. (The ratios, of course, are the cosines of the angles of inclination.)

The drawing is now placed under the tray, and the metal and perspex strips are clamped in positions corresponding to those of the drawing. A suitable potential difference is applied between the electrodes. The ratios of the experimentally derived normal potential gradients are then compared with those obtained graphically. If the ratios agree, then the boundary shape has been found. If they are not similar, the positions of the shapes must be altered on the graph paper, and the procedure repeated until the ratios are similar.

The method is applicable to either cathode or anode shape. The procedure, of course, is simplified if the anode shape is fixed and the cathode shape is to be found. In that case, no alteration of electrode shape on the graph paper is required; the cathode shape need only be altered on the electrolyte tray until its position yields satisfactory agreement between the above ratios.

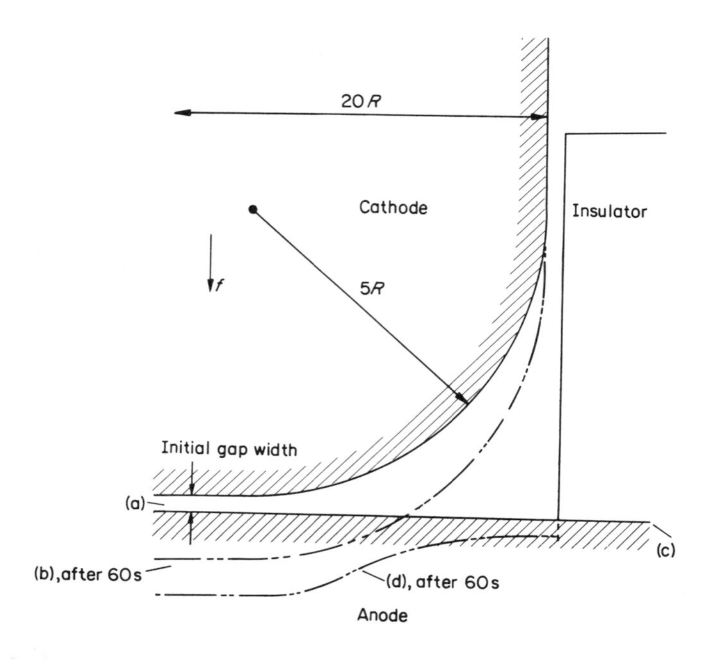

Fig. 7.8 Initial and intermediate positions of cathode [(a) and (b)] and anode [(c) and (d)] (after Kawafune et al. [7])

The same procedures can also be used to find intermediate electrode shapes during the machining time taken to achieve equilibrium conditions. Figure 7.8 shows results from one investigation in which the cathode shape was fixed and the anode was initially flat. The

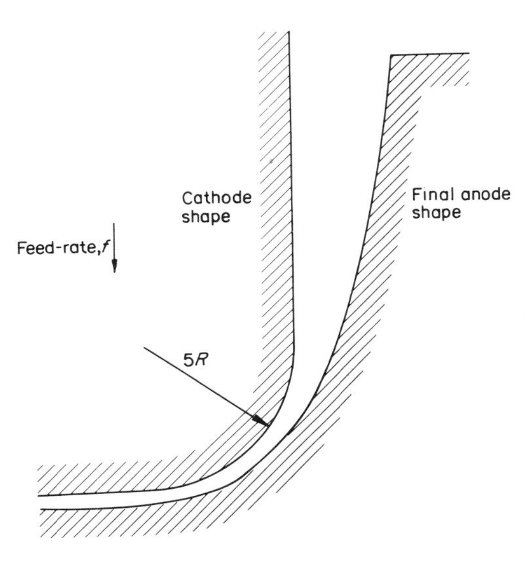

Fig. 7.9 Final anode shape predicted by electrolytic tank analogue (after Kawafune et al. [7])

position occupied by the cathode after a machining time of 60 s is also shown (the cathode feed-rate was 0·002 mm/s.) The resultant anode shape, corresponding to this new position of the cathode, was obtained by the electrolytic tank method, and is also indicated. Repeated applications of the technique yielded the final equilibrium anode shape given in Fig. 7.9. Good agreement was found between the equilibrium anode shape predicted by this method and the form achieved under ECM conditions.

7.3.2 Conducting paper analogue method

The conducting paper method is similar to that of the electrolyte tank, except that the cathode and anode shapes are represented by boundaries outlined in conducting paint on conducting paper. A stabilised variable d.c. power supply is used to supply constant current to the paper, and a voltmeter is used to measure the voltage differences between two probes positioned at appropriate places on the paper. As before, one boundary, the cathode, say, is fixed and the possible shape of the other boundary (the anode) is prepared (Fig. 7.10). The anode shape is also usually sectioned so that the local current density can be measured along it. To simulate the

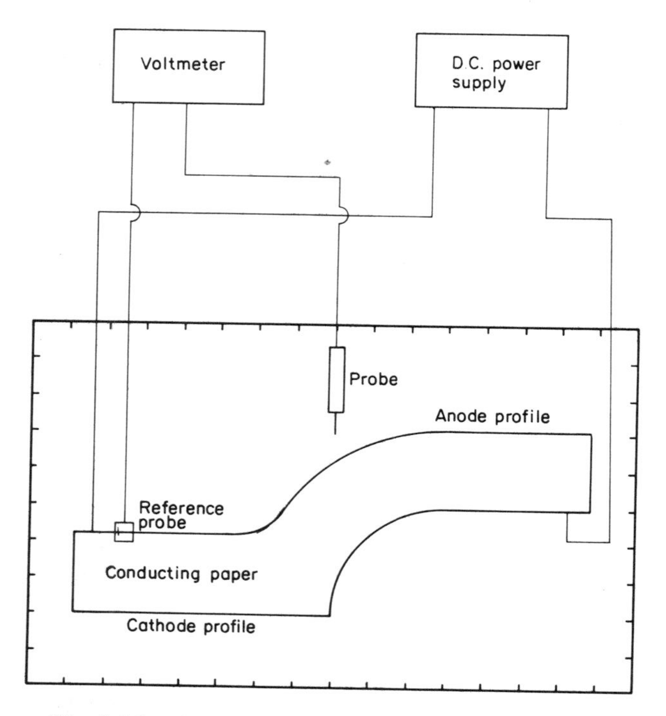

Fig. 7.10 Apparatus for conducting paper analogue

dynamic conditions of ECM, the position of the anode boundary must next be moved through a short distance which corresponds to the change in position after a short time interval, determined by the local ECM conditions (the metal dissolution rate, and hence the current density, and the cathode feed-rate). Such movement can only really be represented by the preparation of a fresh piece of conducting paper for each time interval. Since this procedure is not particularly practical, an alternative device is recommended [8].

Because of the direct ohmic relationship between current and voltage, current distribution lines can be assumed to obey Laplace's equation in the same way as potential lines. Thus the direct analogue, in which potential is constant along the cathode and anode boundaries,

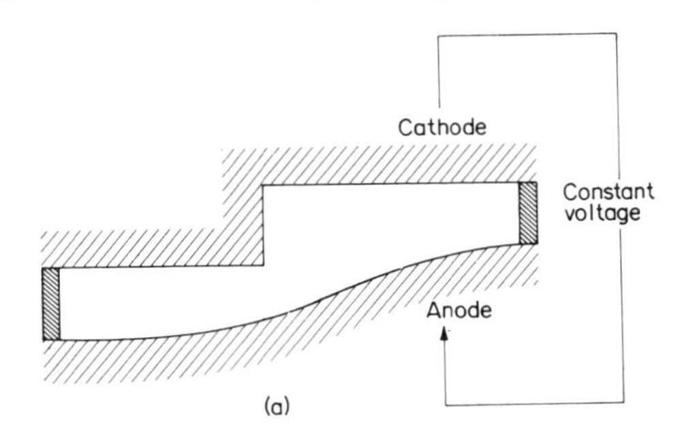

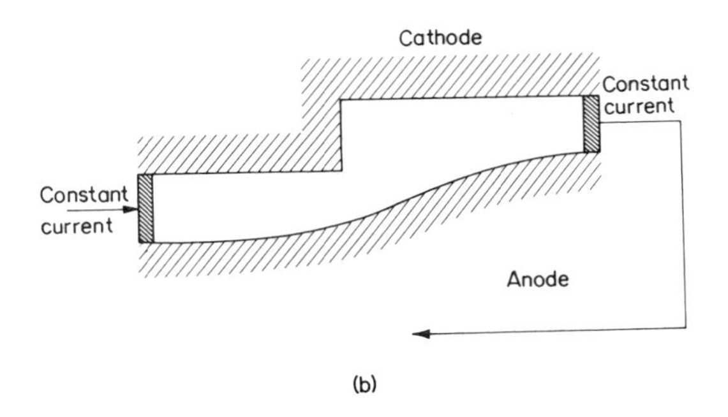

Fig. 7.11 Conducting paper method: (a) direct analogue; (b) 'inverted' analogue (after Tipton [8])

can be replaced by an 'inverse' analogue, in which a constant current is passed through the electrodes painted along the current flow lines (see Fig. 7.11).

The current density J is now represented by the potential gradient $\partial\phi/\partial s$ along the cut edge of the conducting paper, which edge represents the anode surface. This potential gradient is measured from the two probes placed a short distance apart on the edge of the paper.

Consider now Fig. 7.12 which shows conditions at the anode boundary. The vertical vector f represents the feed-rate of the cathode towards the anode. The vector, $\kappa_e(\partial\phi/\partial s)$, which is normal to the anode surface, represents the machining rate and is proportional to the voltage gradient at A. If these velocities are assumed to be constant over time δt, the distances corresponding to these vectors are respectively $f\delta t$ and $(\partial\phi/\partial s)\delta t$. After time δt, the machining causes point A to be transformed to point B. The new anode surface after time δt is now determined by a series of points such as B. The conducting paper is next cut through these points to obtain the new surface. The above procedure is then repeated until no further

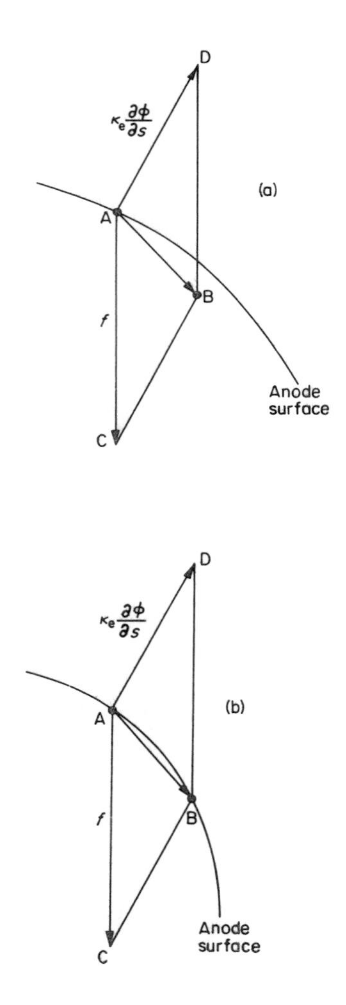

Fig. 7.12 Inverted analogue method: conditions at anode boundary (after Tipton [8])

change in the anode shape is obtained. This final shape is the equilibrium one. Figure 7.13 shows equilibrium anode shapes, predicted by this method, for a stepped cathode with different step sizes [8].

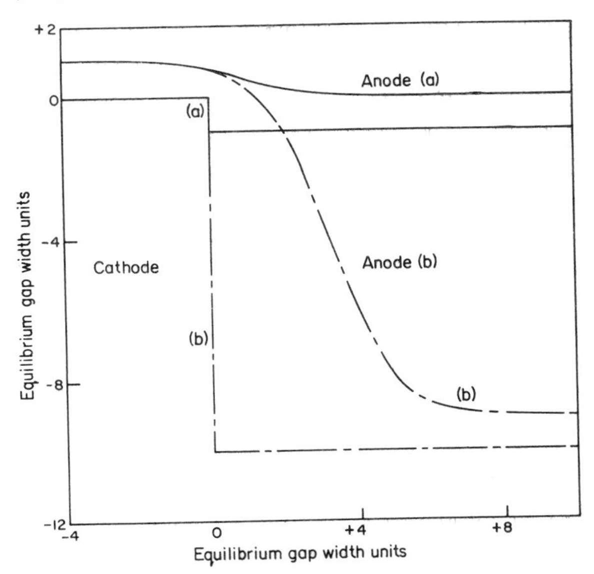

Fig. 7.13 Anode shapes formed with a cathode with step sizes of (a) 1 equilibrium gap width, (b) 10 gap widths (after Tipton [8])

Both analogue methods are tedious to use. In any case, solutions to the field equations for ECM are more easily obtained by means of a digital computer. This approach is discussed in the next section.

7.4 Numerical methods of solution

Numerical methods for solving the shaping problem have the same basis as analogue methods. Suppose that the cathode shape is known and that the anode shape is to be found. An initial, approximate shape, close to the eventual form, for the anode is again chosen, and solutions are sought for the configuration defined by the two electrode shapes and two side walls. Its solution gives the potential at any point between the two electrodes and particularly at the anode. From Ohm's and Faraday's laws the change in shape of the anode boundary, after a small increment of time, can now be found. For these new positions of both electrodes (since the cathode also moves through a distance in the small time interval) Laplace's equation must again be solved. This procedure is repeated until no further appreciable change occurs in the anode shape. The main features of the application to ECM of the well known numerical and computational techniques [9] for tackling this class of problem are now outlined [5, 6, 10].

7.4.1 *Basic method*

Numerical solutions to the problem require specification of the coordinates of both the cathode and proposed anode shapes. These coordinates can be defined in terms of an equi-spaced rectangular

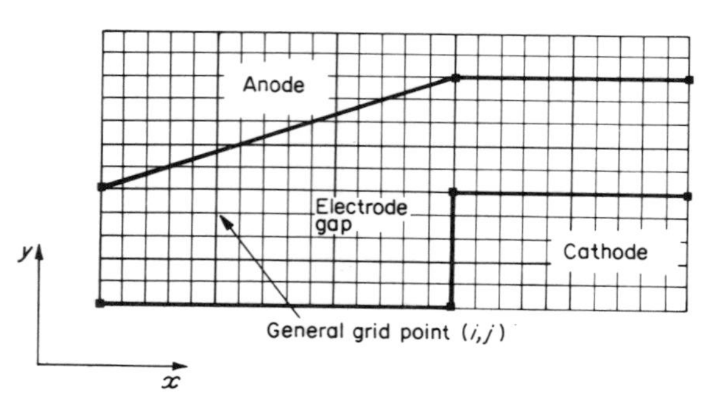

Fig. 7.14 Mesh of grid points defining cathode and anode shapes in terms of coordinates (i,j)

mesh containing a set of grid points of general coordinates (i,j) (Fig. 7.14). A suitable scale, in terms of the mesh spacings, is now chosen for the initial cathode and anode shapes. The size of the scale used depends on the size of computer store available. In turn, the selected scale sets a limit on the electrode dimensions which can be treated. If, as has been recommended [10], five mesh spaces per equilibrium gap unit provide reasonably accurate solutions, then a computer store with, say, 8000 locations limits the mesh area to 320 equivalent units (in terms of the equilibrium gap). That is, the linear dimensions of the electrodes are about 20 equilibrium gap units. Thus for a gap of 0·5 mm this distance is 10 mm.

By this procedure, equipotential boundary lines are obtained which define the electrode forms as well as the electrode gap region in which Laplace's equation is to be solved. (Note: accurate solution to Laplace's equation also requires that the ends of these lines should be terminated in regions where lines of current flow are known, for example, along a plane region of the cathode, or along a line of symmetry parallel to the direction of cathode movement.)

With each grid point is associated some value of the potential within the electrode gap. In addition, the potential values at the grid points on the cathode boundary are made zero, and the values on the anode boundary are set to some large value (e.g. 10 000).

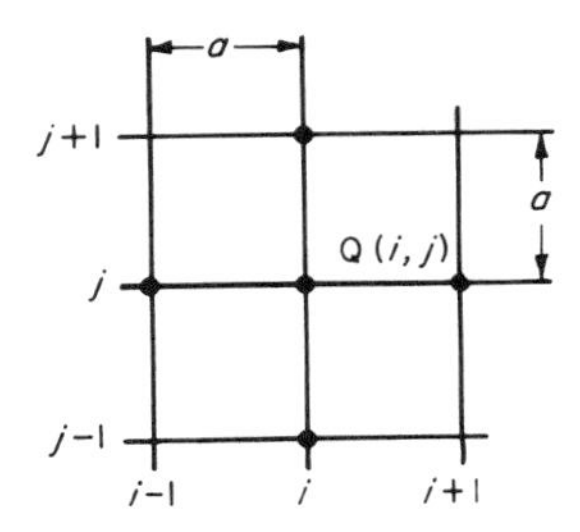

Fig. 7.15 Development of finite-difference method: mesh of side a, with general point Q(i,j)

This same value is applied to the bulk anode within the boundary, so that, as the anode is machined, its fresh boundary takes on the same potential value. Those potential values at grid points within the gap are obtained by linear interpolation along the vertical grid lines between anode and cathode. These values at the grid points must now be progressively adjusted until they satisfy a finite-difference equation corresponding to Laplace's equation in the region between the boundaries.

This method can be explained from Fig. 7.15, which represents a square mesh of side *a*.

The finite-difference equation corresponding to Laplace's equation is

$$\frac{\delta^2 \phi}{\delta x^2} + \frac{\delta^2 \phi}{\delta y^2} = 0 \tag{7.31}$$

in which, given that

$$\frac{\delta \phi}{\delta x} = \frac{\phi(i+1,j) - \phi(i,j)}{a}$$

$$= \frac{\phi(i,j) - \phi(i-1,j)}{a} \tag{7.32}$$

it can be shown that to order (a^2)

$$\frac{\delta^2 \phi}{\delta x^2} = \frac{\dfrac{\phi(i+1,j) - \phi(i,j)}{a} - \dfrac{\phi(i,j) - \phi(i-1,j)}{a}}{a}$$

$$= \frac{\phi(i+1,j) - 2\phi(i,j) + \phi(i-1,j)}{a^2} \tag{7.33}$$

$$\frac{\delta^2 \phi}{\delta y^2} = \frac{\phi(i,j+1) - 2\phi(i,j) + \phi(i,j-1)}{a^2} \tag{7.34}$$

Therefore, the above equation becomes

$$\phi(i+1,j) - 4\phi(i,j) + \phi(i-1,j) + \phi(i,j+1) + \phi(i,j-1) = 0 \tag{7.35}$$

That is, at any point (i,j) in Fig. 7.15, the potential is equal to the average of the potentials at the four neighbouring points.

Next, consider the electrode boundaries, where the potentials are constant. The points on those boundaries do not necessarily lie exactly on mesh points. This means that, for a mesh point adjacent to the boundary, neighbouring points might not have arms of equal length. Accordingly, in the analysis which led up to Equation (7.35) quantities which include the length of the arm (formerly a^2, a constant) now cannot be cancelled out. For such cases, the above equation clearly cannot be used in its present form.

For instance, suppose that the line PR in Fig. 7.16 defines part of an electrode boundary. Whereas, before, the potential could be easily relaxed at the grid point Q, the adjacent points do not now form a regular array for the same operation to be carried out. In this event, it is customary to relax the potential at, say, point S

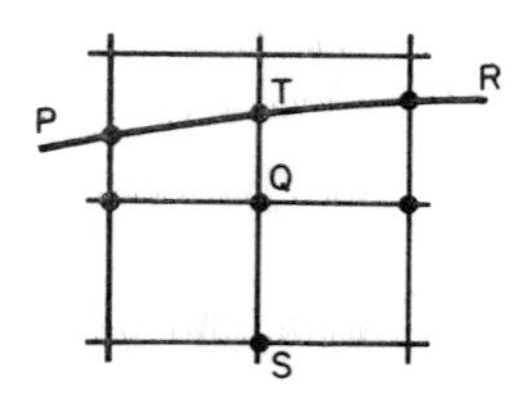

Fig. 7.16 Section of electrode boundary, PTR; long arm ST; potential at Q found by linear interpolation

which has one long arm ST, and then to set the potential at point Q by linear interpolation.

The method of solution now is to adjust repeatedly the potential at each mesh point within the boundaries to a value which is the mean of its four neighbouring points. Eventually, the potential values should approach the correct amounts. During each run, an indication of the deviation of the potential from its correct value at any point can be found from the 'residual' value. This device can be explained from the modified form of Equation (7.35):

$$\phi(i+1,j) - 4\phi(i,j) + \phi(i-1,j) + \phi(i,j+1) + \phi(i,j-1) = R \quad (7.36)$$

where R is the residual, the amount by which the right-hand side of Equation (7.36) differs from zero.

If the potential at any point (i,j) is made the mean of the four neighbouring points, R is zero, or 'relaxed'. In other words, a quantity ($= R/4$) is added to $\phi(i,j)$. The maximum value of the residual gives a measure of the departure of the particular solution from the correct one. The general solution is achieved when the residual becomes smaller than a specified amount. Methods for carrying out this procedure are available elsewhere [6, 9].

The solution so found for these two positions of the electrodes is now used to calculate the normal derivatives of ϕ at the anode, and the anode boundary is now moved by an amount specified by the direction of cathode feed and by the local rate of metal removal. The values of potential previously found for each mesh point now form the initial values for the new solution which is to be found for the fresh position of the boundaries. This procedure is repeated until the changes in the anode boundary become less than the specified amount. This final boundary is the required equilibrium anode shape.

The above procedures can, of course, be used to predict a cathode tool shape for an anode shape which has been specified. But, now the appropriate data specify the known equilibrium anode shape and the probable cathode shape. The relaxation procedure established above is now employed to find the extent by which the normal current density values at the anode have departed from their constant, equilibrium values. Dependent upon the amount of these departures, the cathode boundary is moved, and the relaxation procedure repeated, until the boundary conditions are satisfied. This procedure, however, is by no means easy [6].

7.4.2 *Practical considerations*

In the application of numerical methods, some useful guide-lines have emerged; the machining conditions for the particular metal–electrolyte combination must be studied to provide data on the process variables, for instance the equilibrium gap width and the cathode feed-rate. Experiments of this sort will also indicate the extent of effects due to electrolyte flow, etc., although such effects have not been included in the analysis. Care must also be taken to ensure that the anode workpiece carries sufficient stock to allow the equilibrium shape to be formed to sufficient tolerance.

7.4.3 *Example of numerical methods*

Computer solutions to the shaping problem for a variety of fixed cathode shapes have been prepared by Lawrence [6]. Figure 7.17 shows the numerically predicted equilibrium anode profile produced by a semicircular cathode. The figure also shows the profile found by experiment. The upper numbers along the anode boundary represent a non-dimensionalised form of the measured normal voltage gradient; in comparison, the lower numbers show the voltage gradient calculated from the slope of the boundary.

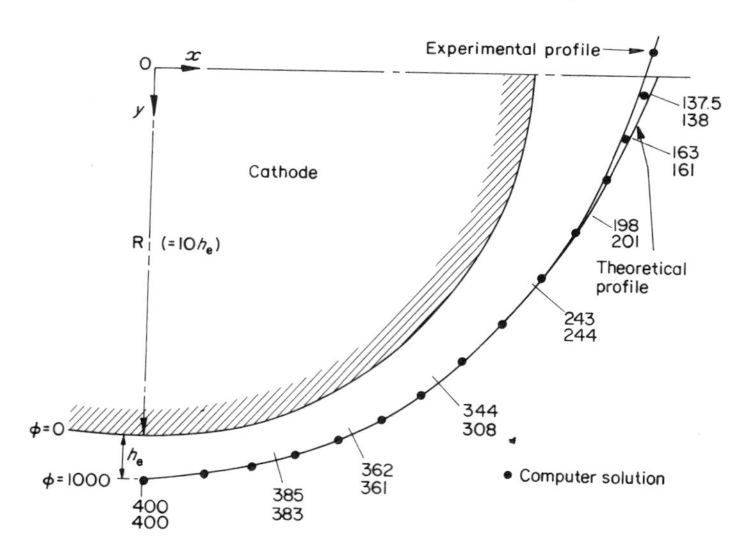

Fig. 7.17 Computer prediction and experimental anode shapes produced by semicircular cathode; Nimonic 90 anode; h_e = 0·76 mm; f = 0·02 mm/s; I = 155 A; V = 15 V; Q = 1·6 x 10^{-3} m^3/s; ρ_e = 1·15 g/cm^3; T = 26°C; pressure at inlet = 750 kN/m^2 (after Lawrence [6])

References

1. Fitz-Gerald, J. M., McGeough, J. A. and Marsh, L. M., *J. Inst. Math. and Applics.* (1969) **5**, 409.
2. Collett, D. E., Hewson-Browne, R. C. and Windle, D. W., *J. Eng. Math.* (1970) **4**, No. 1, 29.
3. Hewson-Browne, R. C., *J. Eng. Math.* (1971) **5**, 233.
4. Tipton, H., Proc. Fifth Mach. Tool Des. Res. Conf., Birmingham, Pergamon Press (1964), 509.
5. Tipton, H., I.E.E. Conference on Electrical Methods of Machining and Forming, London, 1967. Conf. Public. No. 38, p. 48.
6. Lawrence, P., Ph.D. Thesis, Leicester University (submitted).
7. Kawafune, K., Mikoshiba, T. and Noto, K., *Ann. C.I.R.P.* (1968) **16**, 345.
8. Tipton, H., in *Electrochemical Machining* (ed. A. E. DeBarr and D. A. Oliver), MacDonald, London (1968) Ch. 9.
9. Smith, G. D., *Numerical Solutions of Partial Differential Equations*, Oxford University Press, London (1965).
10. Tipton, H., in *Fundamentals of Electrochemical Machining* (ed. C. L. Faust), see Ref. 8, Ch. 4, p. 87.

CHAPTER EIGHT

Applications of Electrochemical Machining

The main attractions of ECM that were outlined in Chapter 1 are its ability to machine very hard metals without causing tool wear and structural damage to the workpiece, and to form complex shapes which are difficult to produce by conventional methods. Its main drawbacks are the high capital cost of the equipment, the cost of tooling, and corrosion problems, which shorten the working life of the equipment. Because of these features, ECM has been used most successfully in aerospace applications where the advantages of using the process, often for a large number of components, outweigh its disadvantages.

Nonetheless, the benefits of ECM are now becoming recognised in other industries, and the process is being used in an increasing number of ways. In some of these applications, the main characteristics of the process are the attraction. In others, ECM has replaced established machining methods because it has been found to give superior results, or to be more economic. Several examples, drawn mainly from these industries, are now discussed which illustrate many of the features of the process studied in the preceding chapters.

8.1 Electrochemical shaping

The application, shown in Fig. 8.1, illustrates the basic principles of ECM, described in Chapter 1. A brass, and occasionally, copper tungsten cathode tool has been used to machine a solid block of stainless steel to the shape required for a knitting machine cam. An applied voltage of 15 to 20 V was applied between the two electrodes,

Fig. 8.1 Electrochemical shaping of stainless steel knitting machine cam (By permission of Healy of Leicester Ltd.)

and the cathode feed-rate was 0·03 mm/s. Under equilibrium machining conditions, the current and current density were 26 A and roughly 78 A/cm^2 respectively, the gap width being 0·25 mm. For these values of the process variables, the shape could be formed on the anode in 270 s.

A requirement of this operation was close tolerance of the machined component – to within 0·1 mm. To achieve this accuracy, a 30% (w/w) $NaNO_3$ electrolyte was chosen. This electrolyte provides good dimensional control without involving the greater hazards met with $NaClO_3$ solution, even though the latter electro-

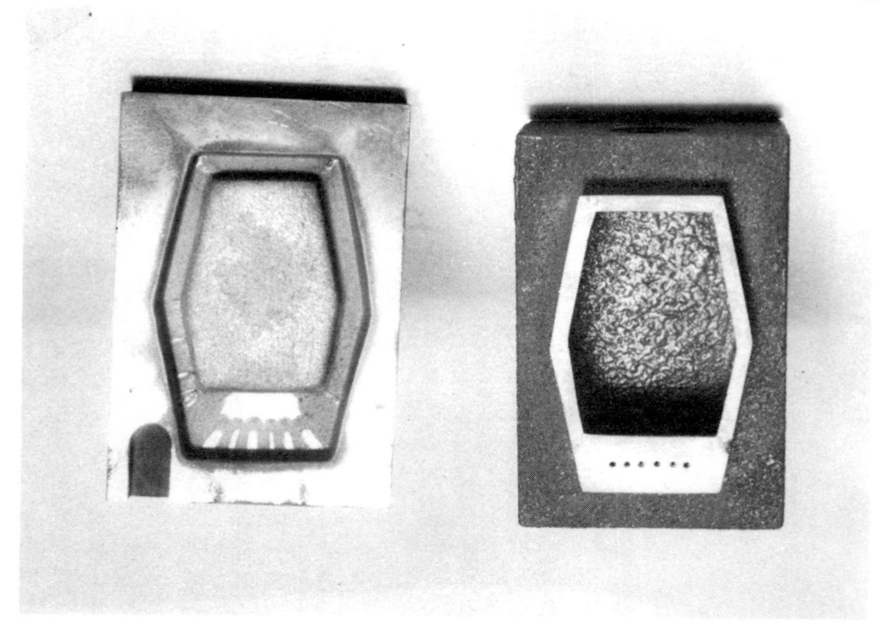

Fig. 8.2 Cathode tool (right) used for electrochemical cavity-sinking preparation of die steel master for rubber moulds; polished streaks on workpiece (left) caused by flow separation (By permission of Healy of Leicester Ltd.)

lyte offers better dimensional control (see Chapter 4). Since hydrogen gas and electrical heating can also upset the accuracy of machining, as discussed in Chapter 5, the electrolyte was maintained at a high inlet temperature of 45°C, and pumped across the breadth (12 mm) of the specimen, the inlet pressure being 1·66 MN/m^2 and the flow-rate (0·76 to 1·14) x 10^{-3} m^3/s.

8.2 Electrolyte flow separation

In the first example, the care taken to ensure a suitable flow path for the electrolyte was pointed out. Good flow conditions are not always readily available in practice, and Fig. 8.2 illustrates a case where poor conditions have led to flow separation and to the termination of machining.

The cathode shape on the right was to be used for a cavity sinking operation at 0·01 mm/s and 15 V to produce the anode shape on the left. The die steel workpiece so formed was then to be used as a master for the manufacture of rubber moulds.

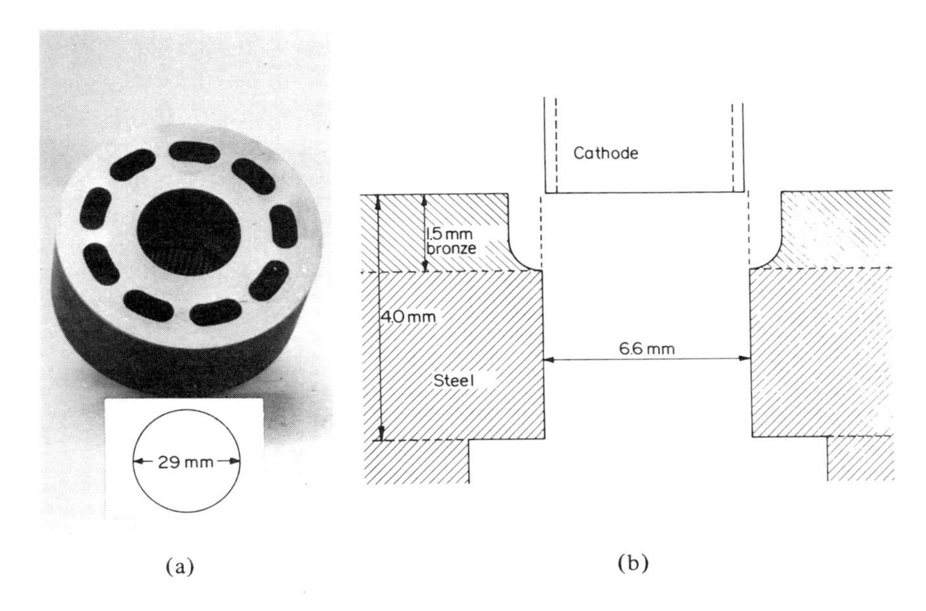

Fig. 8.3(a) Anode workpiece of bronze and steel, used in electrochemical drilling of kidney-shaped slots in pump body (By permission of Lucas Aerospace Co. Ltd., Netherton). (b) Increased overcut through bronze due to difference in machining characteristics of bronze and steel

A 20% (w/w) NaCl solution was introduced through the entry port (top right) at a pressure of $1{\cdot}38$ MN/m^2 and a flow-rate of $(0{\cdot}76$ to $1{\cdot}14) \times 10^{-3}$ m^3/s. The aim was to use the cavity in the cathode as a pressurised chamber for the electrolyte solution during ECM, the outlet for the electrolyte flow being the six holes shown at the foot of the cathode. This configuration, however, led to an uneven flow distribution which caused the polished streaks on the surface of the anode workpiece. Since the current was high – about 700 A – the gap width must have been small, and the irregularities jutting out into the machining gap also caused sparking between the electrodes. The influence of hydrodynamic phenomena (Chapter 2) on surface finish has been described in Chapter 4, and the limitations which they impose on the rate of machining have been discussed in Chapter 5.

8.3 Contrasting metal removal rates

In Chapters 1 and 4, the different rates of machining achieved with different metals were studied in detail. A practical example of the difficulties encountered in machining through two dissimilar materials has occurred in the electrochemical drilling of kidney-shaped slots in a bronze–steel pump body. (Fig. 8.3(a)). In this operation, the head of the cathode tool first met the bronze part of the workpiece, which was 1·5 mm deep (Fig. 8.3(b)). Below the bronze, the tool next encountered steel. The change in material being machined was observed as an increase in equilibrium current. The difference in the machining rates of the two metals was also apparent from the change in overcut, which was considerably less for the steel part of the workpiece. This effect is noticeable in Fig. 8.3(b) as the step at the junction of the bronze and steel.

This example also usefully demonstrates some other aspects of ECM described in Chapters 4, 5 and 7. First, an electrolyte was chosen which gave good dimensional control. The electrolyte was maintained at a constant temperature and pumped at an established flow-rate and pressure in a 'reversed-flow' direction: that is, first down the outer, insulated wall of the cathode, then across the machining face and up through the slot in the tool. This technique diminishes considerably the unwanted effects of hydrogen gas and electrical heating (Chapter 5) by reducing the extent of the flow path over which they can have any influence.

8.4 Mechanical properties of electrochemically machined parts

A benefit of ECM which was explored in Chapters 1 and 4 was that the process does not affect the mechanical properties of machined metals, and moreover, that removal of metal leaves the material in a natural, undistorted state.

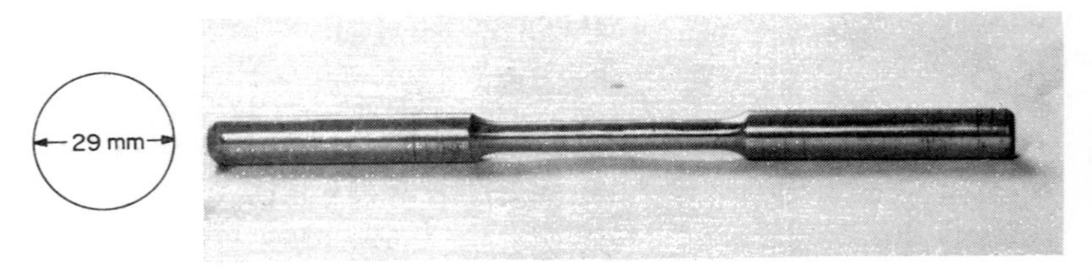

Fig. 8.4 Steel tensile test specimen prepared by electrochemical turning (By permission of Healy of Leicester Ltd.)

This feature has been utilised in the preparation of tensile test specimens for metallurgical work (Fig. 8.4). Here the necked section of the specimen was produced by an electrochemical turning operation in which the steel anode workpiece was rotated on its long axis whilst a cathode tool was moved along it at a rate of 0·006 mm/s. The front gap between the two electrodes was about 0·25 mm, the applied voltage, current, and average current density being 15 V, 100 A, and 78 A/cm^2 respectively. The inlet temperature of the 20% (w/w) NaCl electrolyte was maintained at 35°C, the inlet pressure and flow-rate being 0·69 MN/m^2 and (0·76 to 1·14) x 10^{-3} m^3/s respectively.

8.5 Electrochemical smoothing of irregularities

An increasingly popular application of ECM is anodic smoothing, the main features of which were discussed in Chapter 6. An advantage of this technique in practice is that often no cathode tool movement is needed. Figure 8.5(a) illustrates an example in which surface irregularities had to be removed from the external spline of a shift hub sleeve. Figure 8.5(b) is a schematic diagram of the shape of the stationary cathode tool and configuration of electrodes used for the electrochemical removal of each irregularity whose height was 0·25 mm. The initial gap width between the two electrodes was 0·5 mm. The applied potential difference was 12 V, and with a current of 250 A, about 0·08 mm of metal from the irregularity could be removed in 3 s.

A 30% (w/w) $NaNO_3$ electrolyte at 30°C was used in this work because its good dimensional control restricted ECM mainly to the region of the surface irregularity. The pressure at inlet required for this operation was only about 202 kN/m^2, the flow-rate being about 0·76 x 10^{-3} m^3/s. To minimise the unwanted effects of gas and heating, the electrolyte was caused to flow across the short face of the anode. Figure 8.5(c) shows the configuration of cathode electrodes used for the simultaneous dissolution of all such irregularities on the workpiece. Figure 8.5(d) illustrates the location of both the anode and cathode fixtures.

8.6 Cathode design

Figure 8.6 shows a titanium alloy blade whose complex shape has been produced by ECM. A current of 5000 A (current density

(a)

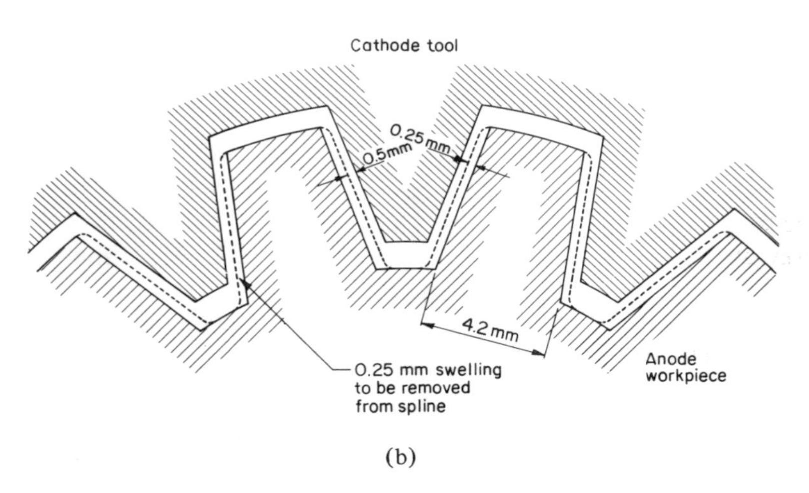

(b)

Fig. 8.5(a) Surface irregularities on external spline of case hardened steel shift hub sleeve (By permission of Turner Manufacturing Co. Ltd., Wolverhampton). (b) Configuration of cathode tool and external spline carrying surface irregularity

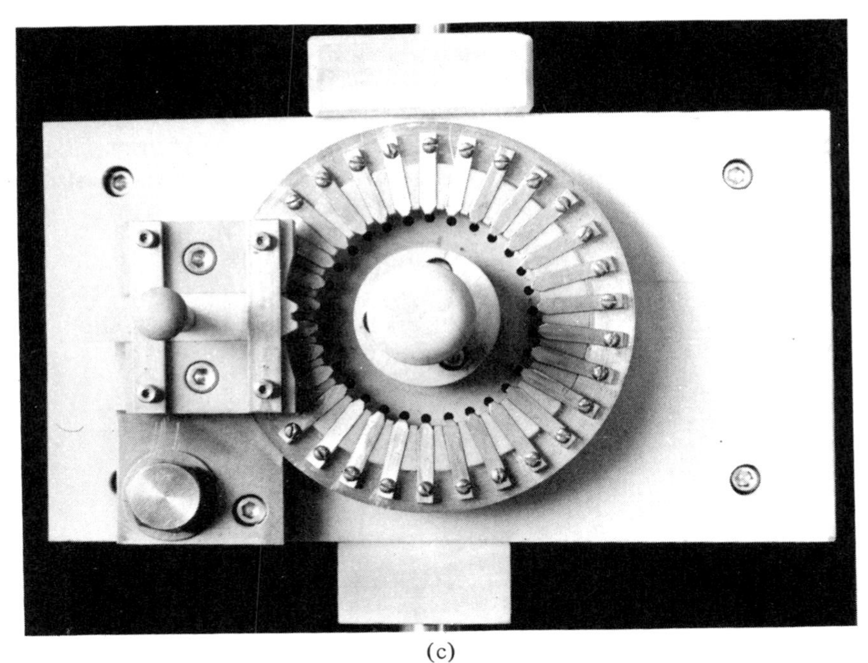

(c)

(d)

Fig. 8.5(c) Multiple cathode tool system for electrochemical dissolution of workpiece irregularities (By permission of Turner Manufacturing Co. Ltd., Wolverhampton). (d) Configuration of multiple cathode tools and anode workpiece for electrochemical removal of surface irregularities (By permission of Turner Manufacturing Co. Ltd., Wolverhampton)

186 A/cm^2) at 10 V was used to machine this tough metal. The cathode feed-rate was 0·05 mm/s and the depth of metal machined was 6·35 mm; the equilibrium (end) gap width was 0·15 mm. Good dimensional control was maintained during machining by use of a

Fig. 8.6 Titanium alloy blade (By permission of Herbert Machine Tools Ltd., Lutterworth)

30% (w/w) $NaNO_3$ electrolyte, the temperature of which at inlet was 40°C. The pressures at inlet and outlet were, respectively, 1·4 MN/m^2 and 340 kN/m^2.

The shape of the cathode tool used to perform this machining operation was obtained by empirical means. However, this anode workpiece is representative of a class of blade shapes which are often much larger, more complex, and have closely defined contours in three dimensions. For these cases, the cathode tool shape cannot easily be found by empirical methods. Instead, computer programs, based on the principles outlined in Chapter 7, are prepared to predict the cathode tool form necessary to produce the anode blade shape. The problem is not a simple one. Not only must the data needed for the program take account of the metal removal rate, which is dependent on the usual process variables and on the choice of a suitable electrolyte, but it should also include some correction for the taper in the machining gap caused by hydrogen gas and electrical heating (Chapter 5). These effects have so far only been predicted for comparatively simple electrode shapes. Much remains to be done to extend these studies to the other electrode configurations encountered in practice.

Bibliography

Wilson, J. F., *Practice and Theory of Electrochemical Machining*, Wiley-Interscience, New York (1971).

Appendix 1 Constituents (percentages) and dimensions of

Metal	Al	Be	Bi	Cd	C	Co	Cr	Cu	Fe	Mg	Mo
Nickel					0·15 max.			0·25 max.	0·4 max.	0·2 max.	
Copper			0·05					99			
Copper								⩾99·9			
Nimonic 75					0·1		19·5	0·5	5·0		
Monel					0·3			31·7	2·5		
Nimonic 80	constituents not given										
Mild steel	Type 1020 (U.S.)								100 (assumed)		
Nickel–chromium	alloy type 56 NiCr MoV7 (Ger)										
Mild steel	constituents not given								100 (assumed)		
Carbon steel					0·06				bal.		
					0·25				bal.		
					0·34				bal.		
					0·52				bal.		
					0·78				bal.		
					0·99				bal.		
					1·26				bal.		
Cast iron					3·03				bal.		
Ni–Cr Steel	Type Carpenter R.D.S. (U.S.)										
Iron				0·03					bal.		
Copper								99·99			
Copper								99·99			
Nickel						1·3					
Stainless steel					0·08		17·81				
Soft iron	Constituents not given										
Steel	Type 1020 (U.S.)								100 (assumed)		
Nickel											
Nickel											
Mild steel	Type 1020 (U.S.)										
Nickel			0·004		0·06	0·23		0·01	0·06	0·08	
Nimonic 80A	1					2	68		5		

principal metals discussed in Chapter 4

Mn	Nb	Ni	P	S	Si	Sn	Ti	U	W	Zn	Other information: Ref. no., shape of specimen face. (C circular, R rectangular, S square, T tubular), dimensions (mm)
0·35 max.		bal.		0·01 max.	0·15 max.						2, C, 33 dia.
				0·05							2, C, 33 dia.
											3, S, 3 x 3
1·0		72·5			1·0		0·4				2, C, 33 dia.
2		63			0·5						3, C, 33 dia.
											6, R, 50 x 12·5
											7, T, 15·9 mm² area
											8
											11, C, 20 mm² area
0·38					0·01						16, C, 13 dia, (polarisation work)
0·76					0·30						
0·81					0·26						16, S, 25 x 25 (ECM work)
0·63					0·18						
0·23					0·64						
0·22					0·23						
0·53		0·1		0·1	2·22						18, C, 33 dia.
											19, T, 7·2 (ins. dia.) (cathode) 19, C, 6·35 dia. (anode)
0·36		0·02	0·016		trace	0·007					20, R, 5 x 20
											21, S, 3 x 3
											21, C, 10³ mm² area
		bal.									5, C, 19·1 dia.
0·88		8·78	0·03	0·17	0·58		0·46				23, R, 12·5 x 50
											27, C, 1·3 dia.
											29, as Ref. 19
											32, R, 12·5 x 22·9
		99·5									33, as Ref. 19
											33, as Ref. 19
0·23		bal.		0·003	0·04						34, C, 6·3 dia.
1		bal.			1		2				34, C, 6·3 dia.

Author Index

Subject Index